RISK MANAGEMENT OF EXISTING CHEMICALS

Proceedings of a Seminar
Conducted

December 8-9, 1983
Washington, D.C.

Sponsored by:

Chemical Manufacturers Association

PUBLISHER'S NOTE

Published by
Government Institutes, Inc.
966 Hungerford Drive, #24
Rockville, MD 20850
U.S.A.

June 1984

Library of Congress Catalog Number: 84-80974
ISBN Number: 0-86587-065-9

Printed in the United States of America

ACKNOWLEDGMENTS

As the seminar coordinators, we would like to express our thanks to Dr. Fred Hoerger, Dr. Leonard Guarraia, Mr. Carl Umland, Dr. Jerry Smith, Dr. Calvin Benning, and Dr. John Behun for their tireless efforts in setting up and moderating the program. We particularly appreciate the program speakers' efforts in bringing their unique expertise and individual perspectives on risk management for chemicals already in commerce. The seminar participants gave freely of their time, expertise, and interest to make this seminar a success. Finally, Pamela Haahr and Kathleen McDermott worked enthusiastically to ensure that the seminar ran smoothly and that no detail escaped their attention. The seminar's success was a tribute to the interest and hard work of all associated with it.

D. Christopher Cathcart
Associate Director
Health, Safety and
Chemical Regulations

Timothy F. O'Leary
Associate Director
Health, Safety and
Chemical Regulations

CONTENTS

FOREWORD

RISK MANAGEMENT OF EXISTING CHEMICALS

J. Ronald Condray
Monsanto Company

The Chemical Regulations Advisory Committee (CRAC) is composed of chemical industry experts in the field of product safety and related disciplines. The group is dedicated to fostering responsible management of toxic substances. CRAC supported passage of the federal Toxic Substances Control Act (TSCA) in 1976 and is committed to promotion of timely and reasonable implementation of the Act. It is in this spirit that we have organized this symposium and look forward to applying the ideas, concepts, and principles that will emerge from the symposium to further the implementation of TSCA.

TSCA provides the Environmental Protection Agency (EPA) with broad authority to require testing, gather information, assess, and regulate chemicals. These activities vary depending on whether or not the chemical in question is a new or existing chemical. Simply stated, an existing chemical is one that is on the TSCA inventory and a new chemical is one that is not.

EPA has made significant strides in implementing a number of provisions of TSCA. There is more to be done, especially in areas of risk assessment and risk management of existing chemicals. These are areas of implementation that create significant challenges to EPA, industry, and other interested groups. There are technical questions such as the translation of animal data to man and modeling techniques. There are policy questions such as safety factors and economic considerations. This symposium will explore in some depth these and other issues that relate to risk assessment and risk management of existing chemicals.

EXECUTIVE SUMMARY

STATE-OF-THE-ART OF RISK MANAGEMENT

Since 1977, the Environmental Protection Agency (EPA) has gradually developed an approach to risk management of existing chemicals under the Toxic Substances Control Act (TSCA). Historically, the agency focused on identifying data gaps and testing needs, and implementing industry-sponsored testing programs under the general TSCA Section 4 testing mandates.

In October 1982, EPA published its TSCA existing chemicals program. The program involves a systematic review of health or environmental effects, risk assessment, the consideration of risk reduction alternatives, and potential regulations. Such regulations could be enforced by other agencies or by other EPA offices under more relevant statutes.

Several other developments in risk management have occurred since the advent of TSCA. Most notable are a continuation of voluntary efforts at testing, research, and risk management practices; advances in analytical chemistry which increase the ability to detect the presence of substances in minute quantities; increased detection of toxicological responses in animal and in vitro systems; and public policy discussion of the roles of science and risk assessment as a part of risk management decisions.

The Chemical Manufacturers Association's "Risk Management of Existing Chemicals" seminar brought together key individuals from several sectors to discuss the state-of-the-art of risk management and the underlying components that influence risk management decisions in both government and industry.

Major conclusions drawn from the seminar include:

- Although there are differences in the rigor and formality of approach, risk management decisions are preceded by: identification, prioritization, and analysis of a hazard; analysis of risk under existing conditions; possible risk reduction options; and selection of the most appropriate risk management option.

- Risk management decisions are made within the framework of statutes or corporate philosophies. Statutory guidance may be definitive, as in the case of the Delaney Amendment, or broad, as with TSCA, which directs a consideration of risk, economics, and social impacts. Scientific uncertainties, public perceptions, and changing societal values are superimposed on the statutory guidance.

- Improvement in the scientific and factual basis for risk assessment is necessary for better risk management decisions and public credibility for those decisions.

PRESENTATION HIGHLIGHTS

CHAPTER 1: SCIENCE AND POLICY IN RISK CONTROL, Dr. J. Clarence Davies

Our understanding of toxic substances recently moved from readily identifiable and measurable "conventional pollutants" to pollutants whose larger numbers and "ubiquitousness in the environment in small amounts" potentially cause chronic adverse health effects. When this happened, assessing toxic risk and making risk management decisions became more complex. These observations are part of the reason Dr. J. Clarence Davies acknowledges, "advances in our ability to detect such substances at low levels is forcing us to come to grips with the implication of a 'zero threshold.' " Furthermore, "pervasive uncertainty" in analyzing risk requires certain kinds of interaction among risk assessors and risk managers and also means that risk assessment cannot be purely a scientific process.

CHAPTER 2: INDUSTRY'S PERSPECTIVE ON HOW RISK MANAGEMENT IS WORKING, Deems Buell

In the fall of 1982, Peat, Marwick, Mitchell & Company surveyed chemical manufacturers' organizational efforts and resource commitments to toxic hazard assessment and control programs. Survey results show most of the 112 chemical manufacturers responding to the survey maintain worker safety and industrial hygiene programs, hazardous chemical assessment programs, product safety programs, transportation safety programs, and hazardous waste disposal programs. Approximately 80% of the responding firms employ health or environmental safety specialists. The survey demonstrates overall attentiveness to self-regulation and compliance with government regulatory statutes.

CHAPTER 3: TOXIC SUBSTANCES CONTROL ACT: NEW ATTITUDES, NEW DIRECTIONS, Dr. John A. Moore

Dr. John A. Moore, Assistant Administrator for Pesticides and Toxic Substances of the Environmental Protection Agency (EPA), raised the issue of public trust in the chemical manufacturers industry and suggests some manufacturers abuse "confidential business information" classification. Questionable use of confidential business information, continues Moore, adds to public perception that the industry has something to hide. While there are legitimate confidentiality claims, questionable claims must be challenged. Moore further supports the development of an "invaluable historical record" of adverse health and environmental effects caused by exposure to chemicals.

CHAPTER 4: QUANTITATIVE RISK ASSESSMENT: STATE-OF-THE-ART FOR CARCINOGENESIS, Dr. Colin N. Park and Dr. Ronald D. Snee

Quantitative risk assessment requires multifaceted scientific analysis. Dr. Colin Park and Dr. Ronald Snee examine mathematical modeling and the many fields of study associated with modeling including "mutagenicity, acute studies in animals, metabolism, chronic studies in animals, epidemiology, [and] route and amount of exposure." Park and Snee discuss various statistical models and review their strengths and weaknesses. They propose models, which include "tumorigenic dose-response information as well as qualitative and quantitative biological factors that affect the estimate of risk." Although mathematical and statistical models provide useful information for risk assessment, models are only part of information vital to risk analysis. Park and Snee conclude by saying, "any extrapolations should be made with great care and only in conjunction with a variety of supporting data. To do otherwise amounts to nothing more than a blind curve-fitting exercise with little predictive value."

CHAPTER 5: GOVERNMENT DATA REQUIREMENTS FOR RISK ASSESSMENT, Dr. Joseph V. Rodricks

Government regulatory risk management decisions already rely heavily on risk assessment and a developing trend indicates this reliance will increase in the future. According to Dr. Joseph Rodricks, however, "risk assessment is a highly uncertain enterprise." Inferences, data gaps, and uncertainties plague assessment processes. Consequently, " 'worst case' assumptions and the resulting estimate of exposure is usually correct...only for a small fraction of the exposed population." The EPA and the Food and Drug Administration (FDA), for example, acknowledge "their estimates of carcinogenic risks are 'upper bound' estimates, and that the true risk will likely be below the upper bound and might even be zero." To avoid unnecessary restrictions caused by data based on upper bound limits that can distort interpretation of realistic risk, Rodricks recommends three remedies: (1) narrow the data gaps for specific substances; (2) improve risk assessment methodologies; and (3) include judgmental, qualitative risk assessment with numerical, quantitative risk estimates.

CHAPTER 6: UTILIZATION OF RISK ASSESSMENT IN CORPORATE RISK MANAGEMENT DECISIONS, Dr. Paul F. Deisler, Jr.

Presently utilized by Shell Oil Company, a generalized risk management decisionmaking approach described by Dr. Paul Deisler had its first application with cancer-causing hazards in cases where existing standards gave inadequate protection. The steps utilized in this system include: (1) identifying and evaluating hazards; (2) establishing degrees of risk according to pre-established risk yardsticks; (3) outlining appropriate actions based on degrees of risk; and (4) implementing management decisions in order to reduce risks. Those who use this approach should recognize it sometimes relies on incomplete and uncertain data and they must further exercise judgment if progress is to be made. There is much room for improvement in the precision of risk assessment as understanding of cancer improves. If those involved face the challenge of assessing and managing toxic substances risks in the face of uncertainty and maintain firm goals for risk reduction, Deisler foresees success in risk reduction efforts.

CHAPTER 7: REMARKS TO THE CHEMICAL MANUFACTURERS ASSOCIATION, Senator David Durenberger

As chairman of the Toxic Substances and Environmental Oversight Subcommittee, Senator David Durenberger suggests that scientists, the chemical manufacturing industry, and government regulatory agencies put "too much emphasis on [risk] assessment and too little on [risk] management." Senator Durenberger intends to review TSCA and other regulatory statutes and to recommend tightening up in some areas and clarification in others. He believes too much information is claimed confidential. This practice, to the detriment of the chemical industry, needlessly denies information to the public. Meanwhile, Senator Durenberger recommends chemical manufacturers continue to clean up waste dump sites, to "encourage the development of effective tools and technologies for risk management, [and] to flood the public record with information about [chemical] products."

CHAPTER 8: INDUSTRIAL VIEWPOINT AND CASE HISTORIES, Dr. Calvin J. Benning

Dr. Calvin Benning introduces two case histories (Chapters 9 and 10) by emphasizing chemical industry problems centering on the slow process of defining goals, producing reliable data, and overcoming mistrust between the public, the government, and the industry.

CHAPTER 9: A CASE HISTORY—PHTHALATES SECTION 4 TEST DATA USE IN REGULATORY CONTROL DECISIONS, Dr. James P. Mieure

Recent regulatory focus on phthalate esters began in 1976 with the inclusion of these substances on the Priority Pollutant Lists. In 1980, National Toxicology Program reports intensified this focus by indicating that the most commonly used phthalate caused tumors in rodents. The EPA, the Interagency Testing Committee, the National Toxicology Program, the Interagency Regulatory Liaison Group, the Consumer Product Safety Commission (CPSC), and CMA's Phthalate Esters Panel promoted subsequent phthalate risk evaluations. The CMA panel is conducting a voluntary testing program accepted by EOA, involving chemical industry and government scientists. Ecological and toxicological data collection is currently underway to determine risk. Ecological toxicity studies include protecting freshwater and saltwater fish and invertebrates. "Acceptable level of intake" becomes the key to human studies and includes oral and other exposure routes. While the CMA's Phthalate Esters Panel prefers to wait for more data prior to issuing risk assessment results, sufficient information is available to make a preliminary risk estimate on di(2-ethylhexyl) phthalate, "a widely used phthalate plasticizer." After analysis of toxicity test results, authorities can agree on regulatory decisions based on available data and social acceptance. The EPA, FDA, and Occupational Safety and Health Administration (OSHA) have separately made statements that "substances having risks below one in a million ought not be subjected to regulation." Phthalate risks above this level may be subject to risk management. Final assessment and risk management decisions await completion of the voluntary testing program.

CHAPTER 10: VINYL CHLORIDE AND TSCA, John T. Barr

The case history of vinyl chloride represents a "bellwether of current regulatory philosophy," according to John T. Barr. Vinyl chloride caught attention about fifty years ago, when it was first marketed as a rubber substitute. Widespread consumer use began about 1950 and now approximately six billion pounds are used annually in the United States. After human health problems developed demonstrating chronic illness associated with workplace exposure to vinyl chloride, "a virtual explosion of research on the chronic toxicity" of this substance erupted. The EPA, FDA, OSHA, and CPSC as well as the Department of Transportation and the U.S. Coast Guard, who regulate transporation of toxic substances, enacted emergency government regulations to reduce exposure. Because of the immediate government regulatory action, risk assessment studies, and scientific advancements, vinyl chloride is seen as a success story in toxic substances risk management.

CHAPTER 11: INDUSTRY'S PERSPECTIVE ON HOW INDUSTRY/GOVERNMENT RISK MANAGEMENT PARTNERSHIP CAN WORK, Dr. John D. Behun

"Protection against unreasonable risk" is the common risk management goal of both the chemical manufacturing industry and EPA. According to Dr. John Behun, EPA assumes regulatory responsibility by taking a five-step approach within its Existing Chemicals Program. Industry, on the other hand, implements internal initiatives to test "a substantial number of chemicals based on internal risk assessments. Some of these have resulted in voluntary or negotiated testing programs with the Agency. In certain instances, the initiatives have gone even beyond the testing programs prescribed by EPA in order to examine additional potential effects." Although TSCA has "triggered" levels of risk assessment and risk management, which Behun acknowledges should not be underestimated, he points out the industry acts voluntarily in these areas rather than responding only to government regulations. Both industry and the government must learn a great deal to achieve effective risk management. One way to promote learning is for government and industry to share information. Not only will these two groups benefit from cooperative efforts, but the public will benefit as well.

CHAPTER 12: TSCA'S ROLE IN THE OVERALL FEDERAL REGULATORY SCHEME, Robert M. Sussman

Robert Sussman outlines the scope of TSCA saying it excludes foods, drugs, cosmetics, and pesticides, which are covered under other statutes. But gray areas exist regarding chemicals which should be covered under TSCA and those which should fall under the jurisdiction of other mandates. Determinations are defined in

Section 9 of TSCA, but Section 9 is a "complex and ambiguous provision that has not yet been clearly interpreted." Furthermore, "informal coordination" between EPA and other agency officials becomes the determining factor in deciding jurisdiction in many cases. This may lead to "conflicting regulatory approaches to essentially identical health or environmental concerns....The absence of clear boundaries between TSCA and other laws frustrates rational planning by industry [and it] undermines the government's own ability to function effectively." Sussman points out that although EPA may wish to preserve flexibility in regulating toxic substances under TSCA, the public would benefit from a "clear set of guidelines."

CHAPTER 13: USE OF QUANTITATIVE RISK ASSESSMENT IN REGULATORY DECISIONMAKING UNDER FEDERAL HEALTH AND SAFETY STATUTES, Peter Barton Hutt

Peter Barton Hutt presents a historical account of attempts to reduce health risks by imposing laws. Until recently, he points out, regulatory action was based on observations of human experiences. Unfortunately, "as long as safety evaluation remained primarily based on direct observation of human experience, safety determinations were almost completely judgmental. There could be no attempt at quantification of risk....With the advent of controlled animal experimentation, however, operational definitions of safety became feasible for the first time." Furthermore, mathematical models were developed which advanced the science of risk assessment until it has become indispensable in risk management decisionmaking. Although most government agencies require "scientifically accepted procedures of risk assessment" and incorporate quantitative risk assessment in regulatory decisions, they recognize the need for acknowledging significant and insignificant risk. Public perception of chemical use, however, is another matter. "Conservative assumption is piled upon conservative assumption, with the result that the 'upper bound risk' that is usually calculated vastly overstates realistic risk potential. Those calculated risks, moreover, are easily misunderstood as 'hard numbers' by the general public." Reeducating the public to abandon "unrealistic expectations" about protection from all exposure is a task the industry and government must face together.

OVERVIEW

ROLE OF RISK ASSESSMENT AT CHEMICAL MANUFACTURERS ASSOCIATION

Dr. Jerry M. Smith
Rohm and Haas Company

The necessity of separating "risk assessment" from "risk management" is a recent revelation. When analytical and toxicological tests were of equal sensitivity, the detection of a chemical suggested a risk and non-detection was equated with no risk. As the analytical chemist developed methods capable of detecting chemicals at nano and pico gram quantities and as the limitations of toxicity testing became known, the "certainty of safe" vanished and the concept of "acceptability of risk" emerged. Furthermore, risk assessment with its uncertainties was recognized as a scientific endeavor, while acceptability of risk was acknowledged as a societal decision.

Risk management, the control of exposure to levels where perceived risks are acceptable, in a free enterprise system resides jointly with the manufacturer, distributor, and public (governmental regulatory agencies are surrogates for the public). However, before sound risk management decisions can be made, sound risk assessments must be made. Sound risk assessments are dependent upon hazard identification, hazard evaluation, exposure identification, and exposure evaluation by competent, experienced scientists using appropriate scientific methods. Unfortunately, scientists cannot determine with certainty the hazard and exposure associated with a given chemical and its uses. Therefore, assumptions and estimations, which are subject to bias, must be made and uncertainties clearly stated.

To minimize bias and provide the best estimate of risk, an understanding of the methods used and the limits of the estimation is required. Through its standing committees, special committees and programs, and liaison with member companies and regulatory agencies, Chemical Manufacturers Association has developed an understanding of the methods and the limits of estimation required for risk assessment. CMA's role in risk assessment and risk management is to provide benefits of its experience.

INTRODUCTION

THE EVOLVING "EXISTING CHEMICALS PROGRAM": AN INTRODUCTION TO THE SEMINAR

Dr. Fred D. Hoerger
The Dow Chemical Company

The Chemical Regulations Advisory Committee's (CRAC) Existing Chemicals Task Group has been a focal point to bring industry expertise on risk management to bear on the challenges of an evolving regulatory program. The Task Group has served as a resource to industry groups which relate to a specific chemical, such as the Chemical Manufacturers Association (CMA) special program panels and similar groups of other trade associations, and has provided information and views to EPA in its development of the program.

In overview, our scenario for a successful risk management program under the Toxic Substances Control Act (TSCA) is to clarify the policy and substance thrusts by the Environmental Protection Agency (EPA), to give examples of industry experience in assessing chemical safety and establishing priorities, to gain a public interest perception of risks, to explore concepts of peer review, and to consider the interrelationship of TSCA to other laws. This scenario candidly acknowledges some of the uncertainties inherent with a new program including advances in scientific knowledge and the variability of political winds.

We believe the seminar comes at a time when the TSCA Existing Chemicals Program is far enough advanced that clarification of the component parts will provide perspective on the places in the process to input information or for constructively voicing concerns. From our Task Group analysis of the Existing Chemicals Program, it seems clear that data and substance will be the basis for many subdecisions leading to final regulatory decision.

During the past two years, significant developments have led to an increasing and more formalized focus on the management of risks associated with chemicals. In March, 1983, the National Academy of Sciences emphasized the distinction between risk assessment and risk management. In June, William Ruckelshaus emphasized the need for the best possible risk assessment as an input for risk management decisions by the agencies. Government policies under both the Carter and Reagan administrations have emphasized the importance of evaluating several options for risk management in terms of effectiveness. Early this year, CMA published survey results of industry practices for the safe manufacture and use of chemical substances. Finally, Congress and the Administration both have spotlighted issues relating to how much risk management is appropriate.

Superimposed on policy considerations has been a strong and increasing stream of testing data on existing chemicals from voluntary industry projects, TSCA Section 4 testing programs, bioassay studies of the National Toxicology Program, and the results of "old" tests collected under Section 8(d) of TSCA. This then is the basis of the TSCA Existing Chemicals Program: the review of toxicity and environmental information, followed by consideration of regulatory control options.

EPA's Existing Chemicals Program is still in the early stages of implementation. As a result, no established or complete "model" for EPA, industry, or public groups to relate to exists.

Until more experience is gained with the program, there will be a degree of anxiety or tension in forecasting the ultimate impact of this program. However, there does seem to be substantial consensus on the EPA premise of carefully evaluating new data, systematically collecting exposure profile information, and the logic of the five phases of the program: hazard identification, problem characterization, information collection and risk assessment, evaluation of risk reduction options, and the imposition of risk management through regulation where necessary. The tiered decision logic parallels the processes involved in companies in establishing their control practices.

We suggest that an ongoing challenge for both industry and EPA will be to identify, on a case-by-case basis, the information relevant to the regulatory process and an efficient approach to its generation and/or collection.

CHAPTER 1

SCIENCE AND POLICY IN RISK CONTROL

Dr. J. Clarence Davies
Conservation Foundation Inc.

INTRODUCTION

One of government's primary functions has been to protect its citizens against risks. Police seek to protect us against risks to life and property; armies against risks of foreign invasion; engineers against risks of floods. Over time, we have asked government to provide safeguards against an increasing number and greater range of risks beyond those involving immediate danger to our health and safety. We have instituted a number of programs to protect us against economic risks as well as against risks to the stability of our environment, to the survival of individual animal species, to natural amenities we consider important, and indeed to the entire range of conditions and activities which contribute to the quality of our life.

In all these cases, some form of risk assessment is necessary before the government takes action. In the past, many risks were so obvious no formal analysis was required beyond a reasonably clear understanding of what the risks were and what was needed to control them. This is seldom the case any longer. Risks associated with an exploding steamboat boiler are much clearer than those associated with the use of a pervasive chemical such as formaldehyde or benzene.

Within the past 15 to 20 years, the increased risk of cancer from environmental pollutants and from chemicals has worried industrial societies. Prior to about 1965, the major hazards from water pollution were considered to be its threats to fish and wildlife and, in the case of air pollution and occupational exposures, "acute" health effects which manifest themselves within a few minutes or at most a few days. Testing of chemicals for their ability to cause cancer, birth defects, genetic changes, or other "chronic" health effects was a rarity, mostly limited to academic laboratories.

However, chronic adverse health effects are now the primary rationale for controlling environmental pollutants. Since the early 1970s, the great majority of proposed federal regulations limiting exposure to chemicals has been based on grounds that the chemical is a carcinogen.

Earlier, it had been generally agreed that if a half dozen air pollutants (the "criteria" pollutants) and a similar number of water pollutants (fecal coliform, suspended solids, Biochemical Oxygen Demand (BOD), nitrogen, and phosphorus) were successfully controlled, there would be no pollution problem. However, once chronic health effects became the focus of concern, the universe of hazardous pollutants exploded. Literally hundreds of chemicals and metals are now or have been candidates for regulatory action. These radically different types of pollutants, the so-called toxic substances, are found in all parts of the environment—air, water, and land—and usually in very small amounts compared to the conventional pollutants.

The characteristics of toxic substances—their large number, their ubiquitousness in the environment in small amounts, and their potential for causing chronic health effects—make more explicit risk assessment an important part of regulatory decision-making. Deciding which chemicals to regulate, at what levels, and where in the environment requires a careful effort to define the various risks involved.

REQUIRED TRADE-OFFS

With so many new and dissimilar risks, it has become increasingly clear that society lacks the administrative and economic capacity to control all of them to the maximum extent. In many cases, there is no way to eliminate the risk entirely without eliminating the activity that causes it. For instance, it appears that many toxic substances have no threshold exposure level below which adverse effects will not occur. Exposure to any level of such substance increases risk. Advances in our ability to detect such substances at very low levels is forcing us to come to grips with the implication of a "zero threshold." We have found that exposure to toxics is widespread and, in fact, inevitable in an industrialized society, and thus zero risk is not an attainable goal. The risk from some individual toxic substances may be reduced to zero, for example by banning them, but eliminating all risks from toxics is impossible.

We are also coming to realize risks are associated with regulatory action as well as inaction. Nuclear power plants, no matter how carefully designed, constructed, and operated, pose small but finite risk; but attempting to eliminate this risk altogether by banning such plants would result in more electricity being generated by coal fired power plants that also create risks. Banning nitrites from food would increase the risk of botulism, or might lead to substitution of another chemical with risks of its own. In these cases, therefore, even eliminating the activity creating the risk would not eliminate risk. It would only result in the substitution of other, potentially more serious risks.

It is also necessary to trade off risks against other societal values such as economic growth. The need for these trade-offs has been reinforced by the poor performance of most western economies in recent years and, at least in the United States, by resentment against government regulation.

For these reasons, then, making trade-offs is a critical element in the process of assessing and controlling risks. Obviously, this increases the amount of information needed in order to make such decisions. But while our information needs are increasing, we actually have less information available about the specific nature and magnitude of risks.

PERVASIVE UNCERTAINTY

Virtually all elements of risk assessment are clouded with uncertainty. The uncertainty is basically of two kinds. First, the various scientific disciplines involved in assessing risk are not sufficiently refined either to explain the mechanisms by which particular causes produce particular effects or to provide quantitative estimates of cause-effect relationships. Second, the data needed to analyze particular risks usually are not available. Typically, we do not know how many people are or will be exposed to a hazard, how much of a toxic substance will be present in a specific part of the environment, how a new technology will actually perform in practice, or how much it would really cost to reduce a given hazard by a given amount.

The pervasive uncertainty about risks and risk reduction has, somewhat ironically, reinforced the importance of risk assessment and intensified the emphasis on it. Much risk assessment methodology involves decision rules for dealing with uncertainty. Since the process relies so heavily on assumptions and guesswork, it is particularly important that all steps be documented and explicit to the maximum extent possible. Only in this way can an assessment's credibility rest on some basis other than the reputation of whoever performed the assessment.

In the face of scientific uncertainty and uncertain data, some interested parties magically invoke the term "risk assessment" in an effort to bring authority and confidence to an uncertain process.[1/] Insofar as it focuses attention on an important field of study and leads to improvements in methodology, such invocation is useful. However, to the extent that it deludes politicians or the public into thinking the analytical process of risk assessment can eliminate the need for more scientific knowledge or better data, it leads to misplaced confidence and false expectations. The use and misuse of the term risk assessment make it particularly important that the term be carefully defined.

ESTABLISHING DEFINITIONS

Risk Assessment

As the popularity of the term risk assessment has increased, varied interpretations of its meaning have proliferated. At one extreme, the term is used to encompass all societal functions related to risk, from identification that a risk exists to implementation of risk reduction measures. At the other extreme, the term has

---

1/ "Cost-benefit analysis" was similarly invoked several years ago.

been limited to the methods used to quantitatively extrapolate human cancer risks from toxicological studies on animals.

I define risk assessment here as the process of determining the adverse consequences that may result from the use of a technology or some other action. The assessment of risk typically includes: (1) an estimate of the probability of the hazard occurring; (2) a determination of the types of hazard posed;[2/] and (3) an estimate of the number of people or things likely to be exposed to the hazard and the number likely to suffer adverse consequences.

When defined in this manner, risk assessment can be used for several different purposes—to establish priorities for further risk assessment or for research; to inform the public about risks; and, as part of the regulatory process, to decide what risks should be regulated and what the content of the regulations should be. A determination of what actions should be taken to control a risk moves the process from risk assessment to "risk management."

Risk Management

Risk management encompasses all activities involved in actually doing something about a risk. First comes a decision on whether any actions are necessary, and if so, what the nature of the actions should be. This decision must be based not only on measuring risk, but also on judging the acceptability of that risk, "a matter of personal and social value judgment."[3/]

In addition, risk management includes implementing the actions decided upon and evaluating their effect.

In short, risk management decisions are always grounded in some sort of risk assessment, although this may be no more than a

---

2/ The use of the terms "hazard" and "risk" is also confused and inconsistent in literature. Generally, in the United States hazard means the probability of an adverse effect, and risk includes the likely magnitude of the effect, i.e., measuring the risk of a chemical includes not only its inherent toxicity (its hazard) but also the number of people likely to be exposed to it. In Europe the terms are usually used in the reverse way. See, for example, The Royal Society, Risk Assessment (London, The Royal Society, 1983), p. 22.

3/ William W. Lowrance, Of Acceptable Risk, (Los Altos, CA: William Kaufman Inc., 1976), p. 8.

decision maker's implicit assumption about the seriousness of the risk.

Deciding which risks will be assessed and which will be controlled first are further steps in the process of dealing with risks. Although such "priority setting" is a critical stage in controlling risk, it is usually not explicitly examined. In discussions of risk control, the implicit assumption is made that the agenda is rational and that all appropriate chemicals are on it. There have been very few examinations of how chemicals or other types of risks actually do get on the government's agenda.

Risk Control

There is no generally accepted term to describe the entire process—priority-setting, risk assessment, and risk management—for dealing with risk. I shall use the term "risk control" for the entire process, although it should be understood that the term is not meant to imply that positive actions should be taken to control all possible risks.

The three elements of risk control are more interactive than sequential. Although a risk must get on the agenda before it is assessed, the agenda is based on at least some informal comparative evaluations of risks. Similarly, although risk management depends on some form of risk assessment, exposure is a key element in risk assessment and the amount of exposure is determined by risk management decisions. Thus, each of the three basic elements of risk control may be repeated several times in varying forms during the course of deciding what to do about a particular risk. A rough and intuitive assessment will put a risk on the agenda for a more elaborate assessment. The more elaborate assessment may result in a decision to reduce the risk, and that decision may in turn result in a still more elaborate and sophisticated assessment.

SCIENCE AND VALUES IN RISK ASSESSMENT

As I have noted, one of the fundamental characteristics of many risk assessments is their pervasive uncertainty. We may be uncertain how often, if ever, the event creating the risk will occur, how serious the hazard is, and how many people will be exposed.

One of the most important weaknesses, and the one most difficult to detect, is that analysts ignore some of the hazards that may arise. This occurs because they overlook hazards, they think hazards are unimportant, or they shy away from them because they are difficult to assess. The way in which a hazard problem is defined often predetermines the conclusion about its seriousness.

For example, the chemical industry, as well as the academic community and government agencies, for many years equated the safety of a chemical with the absence of acute toxic effects. Chronic hazards such as cancer, mutagenicity, or sterility were simply not considered in hazard assessments. This viewpoint is still evident in statements such as, "I've been working with this chemical for 15 years and I've never yet seen any adverse effects."

Another pervasive problem in hazard assessment arises from the inadequacy of scientific knowledge to provide sufficient and unambiguous methods for appraising certain types of hazard, especially the cancer hazards of chemicals. A recent analysis listed 36 steps involved in carcinogenic risk assessment. For more than half these steps there is no consensus among scientists about the right method to use. There is no scientific basis for choosing one over another; choices are essentially based on values or policy. Yet the choice of method will determine the outcome of the assessment in a fundamental way.

These limitations should not be underestimated. Like most human endeavors, hazard assessment is as much art or philosophy as science. However, an explicit and quantitative assessment at least allows these limitations to be perceived and weighed. An intuitive assessment of hazard, based only on the instincts of a scientist, engineer, or decision-maker, is likely to contain even less science and to have even more problems and limitations.

RISK ASSESSMENT'S RELATIONSHIP TO RISK MANAGEMENT

Some interactions between the risk assessment and risk management functions are inescapable. They also are controversial.

Some scholars and practitioners consider a risk management decision to be an integral, inseparable part of risk assessment; others insist that they are totally distinct. To some extent the arguments are based on differences in definition, but there is more to it than that.

Those who believe in separating the two functions note that risk assessment is scientific in nature whereas risk management obviously involves policy considerations. Some of the fallacies in this distinction have been pointed out above. These include policy decisions made in the assessment process, particularly in deciding how to deal with uncertainty; who should make policy decisions; whether or not they should be open and explicit or hidden in the analysis; and whether or not they should be made consistently.

The types of policy considerations involved in risk assessment are, however, different in several respects from those involved in risk management. Both hinge on basic policy views about how cautious or conservative to be about risk. But risk assessment involves a choice about the general methodology to be used for assessing risks. Once this choice has been made, it can be consistently applied to all assessments. In contrast, many of the policy considerations involved in risk management are unique to a particular decision. The political, economic, and social costs and benefits differ in each case.

Those who espouse the separation of risk assessment and risk management are concerned that political, social, and economic considerations in a specific management decision will improperly influence the assessment process. They worry, for example, that a zealous regulator who wants to ban a chemical will apply pressure on assessors to exaggerate the risk of the chemical or will distort the findings of the assessment. This is a valid concern, although the extent to which such abuses actually occur is a matter of dispute. In the absence of policy guidelines for the conduct of risk assessments, it is likely that a smart regulator can make a risk assessment come out any way he wants—and do so without violating the tenets of acceptable science.

There are, however, other ways in which risk management policy influences the risk assessment process and reasons why some aspects of the risk assessment process should be closely related to risk management. The criteria and analytical procedures used in a risk management process will help dictate the type of risk assessment that is needed—as well as how closely the risk assessment process should be tied to the management process. If a management decision is based on technology alone, there is little reason to produce sophisticated risk assessments and a virtually complete separation between the two functions should cause few problems.

The more the management criteria emphasize "reasonable" risk, however, the more reason for closer integration. For example, regulatory options to reduce the risk of a chemical short of a total ban will involve making choices about the number of people exposed to the chemical, the length of time they are exposed, and/or the amount of the chemical to which they are exposed. Each of these choices will change the assessment of the risk posed by the chemical. Thus a high degree of interaction between risk managers and risk assessors in a regulatory agency will be necessary if management options are to be adequately evaluated.

BIBLIOGRAPHY

Lowrance, William W. Of Acceptable Risk. (Los Altos, CA: William Kaufman Inc., 1976).

The Royal Society. Risk Assessment. (London: The Royal Society, 1983).

CHAPTER 2

INDUSTRY'S PERSPECTIVE ON HOW RISK MANAGEMENT IS WORKING[1/]

Deems Buell
Peat, Marwick, Mitchell & Co.

INTRODUCTION

During the past decade, the United States chemical industry has been the subject of intense public examination. Public advocacy groups, politicians, the media, and industry critics have all expressed concern regarding the risks presented by chemicals to health and the environment.

To assess the industry's level of effort in reducing unreasonable risks, the Chemical Manufacturers Association (CMA) commissioned a survey of the industry, conducted by Peat, Marwick, Mitchell & Co. in the fall of 1982. The survey in particular assessed chemical manufacturers' organizational and resource commitment to:

- testing chemical substances and products;
- assessing hazards;
- collecting and transmitting information about chemicals; and
- maintaining strict hazard control programs.

1/ Preparation of this paper is based on "An Industry Survey of Chemical Company Activities to Reduce Unreasonable Risk to Health and the Environment," Peat, Marwick, Mitchell & Co., February 11, 1983.

The survey, a 22-page questionnaire developed by industry experts, was mailed to chemical manufacturers in all areas of the industry, except pharmaceutical manufacturers. The mailing was not limited to CMA member companies, but included companies enrolled in several other chemical trade associations. The 112 companies that returned completed questionnaires reported total 1981 U.S. chemical sales of $78 billion and total employment of 474,000 persons. They represent approximately 56 percent of the industry in terms of sales and 52 percent of the industry in terms of employment. Survey respondents represent a wide variation in company sizes. Eleven had 1981 chemical sales of more than $2 billion, while 37 had sales of less than $50 million. Companies responding to the survey deal with a total of 115,500 chemical products.

SURVEY RESULTS

Health and Environmental Policies

The survey indicates most chemical companies have implemented formal policies that address health and environmental concerns. Virtually all survey respondents (98 percent) have formal policies addressing worker safety concerns; 94 percent have formal policies regarding medical concerns; 91 percent cite formal policies for product safety; and 90 percent have formal policies for hazardous waste concerns. Formal environmental policies are in place for 89 percent of those firms answering the survey; 88 percent have transportation policies; and 84 percent have formal policies addressing disaster plans.

The survey reveals a high degree of upper management involvement in the establishment of these formal policies. In 73 percent of the companies responding to the survey, executives at the level of corporate vice-president and higher are routinely involved in formulating these policies. Similarly, management is almost always involved in completing the compliance audits and reviews with which companies ensure they are following health and environmental policies and regulations.

Several respondents cite specific examples of upper management involvement. For example, two companies report that the environmental technical staff reports directly to an executive vice-president. Several report they have established high-level review committees, comprised of corporate officers and board members, to develop policy and assess company performance in areas of environment and health.

In addition, companies frequently cite certain management policies to illustrate their high level of concern and involvement in health and environmental issues. These policies include:

- encouraging technical publications;
- stopping production to avoid or minimize health or environmental problems;
- encouraging participation in professional association committees, in outside environmental organizations, and with government agencies;
- using only approved hazardous waste disposal companies;
- providing occupational health manuals and proper employee safety equipment for all plants; and
- maintaining open communication with the Occupational Safety and Health Administration (OSHA) and the National Institute of Occupational Safety and Health (NIOSH).

Staff Resource Commitments to Health and Environmental Concerns

Nearly 80 percent of the companies surveyed have one or more groups whose sole function is dealing with health or environmental matters. Responding companies employ a total of 8,700 health and environmental specialists, roughly one in every 55 employees. Almost 5,500 of these specialists deal with worker safety, health, and water pollution control. An additional 877 are responsible for hazardous waste management. Smaller numbers of specialists are concerned with air pollution control (650), product safety (574), transportation safety (218), legal issues (197), and labeling (138). The firms report that, of the 8,700 health and environmental specialists, 1,166 are environmental engineers. Almost ten percent (837) of the specialists are either physicians (257) or nurses (580). Toxicologists number 497, and there are a reported 491 industrial hygienists. The respondent firms have 38 epidemiologists, 28 pathologists, and nine immunologists on staff.

Thus the survey shows that chemical manufacturers devote, on the average, approximately two percent of staff resources specifically to health and environmental concerns. What must be assessed next is the degree to which chemical companies are utilizing their staff resources to (a) identify chemical hazards and (b) comply with the reporting and regulatory requirements and procedures promulgated by the Toxic Substances Control Act (TSCA).

Assessment Programs

Survey findings indicate that nearly nine of ten chemical manufacturing companies have in place a program for assessing chemical hazards. Of these companies, 85 percent perform chemical hazard assessments on new products, and 76 percent on new processes, while 57 percent conduct such assessments whenever a formulation change is effected. In answer to a survey query regarding chemical hazard assessments for existing products or processes, 41 percent indicate assessments are performed on a routine basis. Particular events trigger assessments on a much more frequent basis: new data, 94 percent of the time; new regulatory requirements, 92 percent; employee concern, 81 percent; consumer concern, 78 percent; process change, 73 percent; and new product use, 63 percent. Of the responding companies, 82 percent conduct formal environmental health and safety reviews for the design of new plants or equipment projects; 51 percent of the firms have chemical hazard assessment committees.

Premanufacture Reporting

TSCA places affirmative reporting and regulatory responsibilities on the chemical manufacturers. Under the requirements of Section 5 of the Act, chemical manufacturers must notify the Environmental Protection Agency (EPA) 90 days before they manufacture or process a new chemical or make significant changes in the use of an existing chemical. The company submits the premanufacture notification (PMN), describing the chemical, its anticipated use, and all available health and safety data. The EPA then has 90 days during which to examine the data and review the health and environmental effects of the chemical before it is introduced commercially. The EPA is authorized to require further information or testing of new chemicals and to restrict manufacture or use where the new chemical is believed to present risk of injury.

Of those firms responding to the survey, 84 percent have a procedure for determining that a PMN is required. However, the impact of a PMN may differ depending on company size and the products involved; 18 percent of respondents indicate they have declined to manufacture one or more products due to PMNs. Ten firms provide explanations; almost all cite the time and expenses involved in PMN testing as prohibitive when compared to the market value of the product.

Testing and Testing Procedures

Section 4 of TSCA outlines requirements for the testing of chemical substances and mixtures to determine whether the chemical poses a threat to health or the environment. Over three-fourths (78 percent) of the manufacturers responding to the survey indicate they routinely perform toxicity tests to evaluate the health or environmental effects of chemicals, and the mean year for initiation of such testing was 1963. In 1981, the 112 companies in the survey tested over 4,400 chemicals. In the same year, the surveyed companies committed a total of 1,200 full-time staff and an expenditure of $138 million for the toxicity testing of new and existing products. The total replacement value of the toxicity testing facilities of surveyed companies in 1981 was about $255 million.

Survey responses reflect significant changes in testing procedures over the past ten years. For example, several companies altered their organizational structure to designate groups or individuals with environmental, health, and safety responsibilities. In addition, many increased staffing for toxicological testing and made major investments in testing facilities and equipment. Perhaps the most important improvements have been in the increased numbers of tests and greater sophistication of the tests used, emphasizing a biomedical approach and using more advanced medical techniques. Finally, improvements in recordkeeping and information collection, particularly through the use of computers and information monitoring, have dramatically changed the testing programs in the past decade.

Toxicology Reviews and Reporting Procedures

Section 8 of TSCA authorizes EPA to collect company information relating to the health or environmental effects of chemicals. Under Section 8(e) of the Act, a firm must file a report with EPA informing the agency that a chemical may present a substantial risk to health or the environment. Of the companies responding to the survey, 74 percent state that they have a company procedure in place for considering 8(e) reports. Of these firms, 90 percent cite toxicology as a source for information leading to an 8(e) review. Employees, customer/consumer reports, research, and data from suppliers are other sources of information (with a response frequency of between 70 and 78 percent). For each firm, between six and seven people are normally involved in an 8(e) review procedure. Slightly more than half (51 percent) of the manufacturers indicate they have filed at least one 8(e) report.

Information Dissemination—Employees and Customers: Information must not only be collected; it must be assessed by the chemical manufacturers and disseminated to both employees and customers. Virtually all firms responding to the survey (99 percent) report they use Material Safety Data Sheets (MSDSs), which summarize all the information known about a chemical. Ninety percent of those firms using MSDSs send them to customers on request, and more than half (60 percent) of the companies report they routinely send MSDSs when information on the chemical is updated. Other methods of MSDS dissemination to customers are inclusion with samples or customer information packets, with contracts, and via periodic mailings using a computerized system.

On average, more than 90 percent of the respondents' chemical products were covered by MSDSs in 1981. Virtually all the companies (97 percent) report they review and update the information sheets as new information becomes available. The average period between reviews is two years. Nearly all responding companies indicate that they maintain MSDS files and that the files are always available to company personnel. About two-thirds (65 percent) report that files are also available to the public. The survey report notes, however, that "a number of the companies that make the files available to the public limit the release of this type of information to people with 'a legitimate need to know.'"

Information Dissemination—Labeling: The second most frequently reported means of disseminating production information was through labeling. In addition to the product or trade name and the name and address of the manufacturer, most labels identify precautions for use (87 percent of respondents so indicating), hazard information (86 percent), emergency procedures (72 percent), code number (68 percent), and the common name (66 percent) and chemical name (64 percent) of the product. Container disposal information and a telephone contact are also provided, albeit at a lesser frequency (41 percent and 34 percent, respectively).

Information Dissemination—Records Maintenance: Virtually all survey respondents formally maintain, at either a corporate, divisional, or local level, lists of chemicals produced and chemicals used (99 percent and 97 percent, respectively). A slightly lesser number (95 percent) report formal maintenance of medical records, while 92 percent formally maintain work site exposure records. Customer complaints are maintained by 86 percent and employee complaints by 85 percent. Nearly all indicate they have established procedures for responding to health and safety questions raised by both employees and customers.

All companies surveyed (100 percent) state they maintain complete lists of products; almost all (98 percent) maintain lists of raw materials and segregated intermediates. Ninety percent of the companies keep lists of solid wastes, while 87 percent maintain lists of effluents and 83 percent, lists of emissions.

Almost all firms use these lists to maintain internal control (98 percent) and meet regulatory requirements (94 percent). A lesser amount (86 percent) utilize the lists for health and safety research purposes.

Information Dissemination—Training and Industrial Hygiene Programs: Another frequently reported method of disseminating chemical hazard information is through training of chemical plant employees. Almost all manufacturers responding to the survey provide training to both hourly and supervisory plant employees in the areas of handling of materials, chemical hazards, and workplace safety. Supervisors receive emergency response training from 96 percent of respondent firms, and in regulatory compliance from 90 percent. In contrast, hourly employees receive emergency response training from 86 percent of manufacturers and training in regulatory compliance from only 64 percent of respondent firms. Both supervisory and hourly employees receive training in off-the-job safety procedures from 63 percent of the respondents.

Nine of ten responding companies have an industrial hygiene program designed to protect workers from chemical hazards. These programs address, in order of frequency cited, manufacturing (100 percent), processing (94 percent), maintenance (87 percent), laboratory (87 percent), packaging (82 percent), incoming raw materials (77 percent), storage (72 percent), product use (70 percent), disposal (62 percent), and transportation (60 percent).

To monitor the health effects of the chemical manufacturing process, 86 percent of the companies surveyed periodically review the medical histories of either all (39 percent) or at least selected (47 percent) in-plant and out-of-plant employees. Approximately one-half of companies performing periodic reviews select those employees who are exposed or potentially exposed to hazardous chemicals. Other in-plant personnel reviewed include manufacturing and production workers, chemical operators, lab personnel, waste handlers, shipping and receiving workers, supervisors, and ex-production workers. Out-of-plant employees cited include sales personnel, management, and truck drivers.

An important element of hazard control is the internal compliance audit program by which companies ensure they are following in-place health and environmental policies. The survey showed 76 percent of the responding companies have a compliance audit or review program. On average, health compliance audits for responding companies were begun in 1975, and environmental compliance audits were begun in 1978. Compliance audits are performed approximately every 14 months, usually by personnel with either technical or management skills. Several companies report they perform other audits besides health and environmental compliance audits, such as audits of safety, industrial hygiene, and transportation, as well as site assessment, quality control, and Food and Drug Administration (FDA) compliance.

Transportation Safety

Another important facet of hazard control is transportation safety. Almost all survey respondents (96 percent) indicate they train their personnel to understand transportation procedures manual covering procedures for loading/unloading, handling, packaging, labeling, placarding, inspection, and spills or other emergencies. Many of the manuals also contain Department of Transportation (DOT) standards and Federal Motor Carrier Safety Regulations, hazardous substance listings, and hazardous waste transportation rules. More than half (64 percent) provide precautionary information beyond DOT standards for identifying the hazardous nature of chemicals in transit. In transporting chemicals and products, virtually all respondents (98 percent) use labels, and most also provide placards and written and verbal instructions. Three-fourths of the companies (76 percent) list their products with CHEMTREC, a CMA/Department of Transportation emergency hotline, and half maintain their own transportation emergency hotlines.

Hazardous Waste Disposal

A final survey area requests data on waste disposal practices. Many respondents (74 percent) indicate they have increased their utilization of recycling as a means of handling chemical wastes. A sizeable number of firms (39 prcent) are using more off-site landfill in recent years than previously, although many firms (30 percent) report they are using off-site landfills less in recent years than previously. Many manufacturers have in recent years increased their use of incineration both off-site (43 percent) and on-site (36 percent); however, many firms also report they have not recently relied upon incineration at all (30 percent for off-site, 48 percent for on-site). Most survey respondents indicate they are either using less on-site landfill than in the past or none at all (84 percent), off-site storage (88 percent), and on-site storage (60 percent).

SUMMARY AND CONCLUSIONS

The survey demonstrates that most U.S. chemical manufacturers have in-place programs demonstrating organizational and resource commitment to diminishing risks posed by their products and manufacturing procedures. In particular, the survey shows that:

- Most chemical companies have implemented formal policies addressing health and environmental concerns with a high number (90 percent or above of respondents) having programs for worker safety concerns, medical concerns, product safety, and hazardous waste problems. Many of these firms involve upper management in the establishment and implementation of these programs.

- Many chemical manufacturers have one or more groups whose sole function is dealing with health or environmental matters. The firms answering the survey report that approximately one of every 55 employees is a health or environmental specialist, the majority of these dealing with worker safety, health, and water pollution control.

- Almost 90 percent of responding manufacturers have in place a program for assessing chemical hazards. Most of these firms (85 percent) perform chemical hazard assessments on new products and three-quarters (76 percent) on new processes. Forty-one percent of firms perform chemical hazard assessments for existing products on a routine basis. More frequent chemical hazard assessments are prompted by occurrence of particular events, such as new data, new regulatory requirements, or employee or consumer concern.

- Many firms (84 percent) have procedures for determining that a PMN is required. Almost one in five firms (18 percent) report they have declined to manufacture one or more products due to the burdens of cost and time imposed upon them by the PMN procedure. Over three-fourths of the manufacturers routinely perform toxicity tests, utilizing 1,200 full-time staff and $138 million.

- Almost three-fourths (74 percent) of respondents have a company procedure for considering reports under section 8(e) of TSCA, which requires firms to file a report informing the EPA that a chemical may present a substantial health or environmental risk.

- Almost all firms (99 percent) utilize MSDSs, which summarize all information known about a chemical. Ninety-six percent of these firms send MSDSs to a customer on request, and more than half (60 percent) routinely send MSDSs when chemical information is updated. More than 90 percent of the respondents' chemical products were covered by MSDSs in 1981. Other often-cited methods of disseminating production information were through labeling and through training of chemical plant employees. Most responding manufacturers have an in-plant industrial hygiene program. Many (86 percent) also systematically review the medical histories of all or selected employees.

- Finally, the survey shows an increased use of recycling as a method for handling chemical wastes. Most respondents report they are using disposal methods such as on-site landfill, off-site storage, and on-site storage less or none at all.

BIBLIOGRAPHY

Peat, Marwick, Mitchell & Co. "An Industry Survey of Chemical Company Activities to Reduce Unreasonable Risk to Health and the Environment." February 11, 1983.

CHAPTER 3

TOXIC SUBSTANCES CONTROL ACT: NEW ATTITUDES, NEW DIRECTIONS

Dr. John A. Moore
U. S. Environmental Protection Agency

INTRODUCTION

In November, I sat before the Senate Committee on Environment and Public Works. At this, my confirmation hearing, I testified that I considered the one key element to the successful execution of my duties as Environmental Protection Agency Assistant Administrator for Pesticides and Toxic Substances to be "trust." By that, I meant the process of maintaining public faith in the government's ability to safely, efficiently, and ethically regulate chemical products. As you know, the Toxic Substances Control Act (TSCA) regulates chemicals for the sake of the environment and the health of the public. I am not a clairvoyant, nor do I need to be to perceive that you in the industry and we at EPA need to realize gains in the magnitude of the American people's trust.

GAINING PUBLIC TRUST

Think of the two words that very often are used by the lay public when it addresses your industry and they are most likely to be "hazardous" and "toxic." And who can blame the public when there is doubt about the plastic nipple on a baby's bottle. When they wonder what else is in a loaf of bread. When the materials which comprise their homes are suddenly suspect. Telling them about the bounty that chemistry has brought them just won't do—not when they suspect it contains unknown time bombs. The key point is that in an environment of distrust the reaction to legitimate information or misinformation is the same: it is often irrational acts.

Not long ago William Ruckelshaus stated he had no doubt that EPA will in some circumstances require expenditures that show an attempt to control very low risks. This will be done because society has told us to pay the price as a sort of insurance. In a certain sense, the actual, quantifiable risk reduction we obtain thereby is beside the point. We are really buying freedom from fear, and most Americans are willing to pay a reasonable price to obtain it.

The antidote for fear is knowledge and the cost of "reasonable price" that Mr. Ruckelshaus alludes to is inversely related to that knowledge. The chemical industry can make great strides toward improved public trust by reexamining its attitude toward confidentiality of health, safety, and environmental data. Many companies have a laudable policy of classifying as confidential business information (CBI) only that information that is necessary to protect a trade secret, but I question the policies of others. Some classify information as CBI to protect it from the public eye and public scrutiny rather than protect it from the eyes of competitors. This practice must stop. I refuse to believe it is a policy that won't work, because some of the most successful companies in the business have been practicing it for years. I plan to ask the Administrator's Toxic Substances Advisory Committee to review this issue and give me guidance.

An important tool under TSCA is the Section 8 authority to gather a broad range of information from your industry about such things as production, use, exposure, and unpublished health and safety studies. Data generated under Section 8 rules will be shared with the regions, other federal agencies, and the public to the extent possible while protecting legitimate CBI claims. As I indicated previously, we intend to continue to challenge questionable CBI claims, especially those relating to health and safety studies.

A significant achievement in the Section 8 area is the recent promulgation of the Section 8 rule which requires manufacturers and processors of chemicals to keep records of significant adverse reactions to health and the environment alleged to have been caused by chemicals. A significant outreach seminar took place in Washington during November 1983 to make companies, labor unions, public interest groups, and the general public aware of the rule's provisions. We feel that the rule will serve as an important tool to alert corporate, labor, and government officials to potential chemical problems, as well as create an invaluable historical record of alleged problems. It will enable industry and EPA to identify patterns of adverse effects which might otherwise go unnoticed for a long time.

CHAPTER 4

QUANTITATIVE RISK ASSESSMENT: STATE-OF-THE-ART FOR CARCINOGENESIS[1/]

Dr. Colin N. Park
Dow Chemical U.S.A.

Dr. Ronald D. Snee
E. I. du Pont de Nemours & Company

ABSTRACT

A critical evaluation of the use of statistical models in carcinogenic risk assessment is made with emphasis on the strengths and weaknesses of current practice. The objective is to bring together information from the fields of toxicology and statistics to develop a sound scientific basis for making risk decisions. It is emphasized that risk assessment is a complex, multifaceted process that is not easily quantified and, at present, must be based on qualitative as well as quantitative information. It is evident that dose-response modeling of tumor incidence data from animal studies

---

1/ This article was first published in the July/August 1983 issue of Fundamental and Applied Toxicology. It is reprinted here with the kind permission of that journal, which is the copyright holder.

is of limited value in estimating human risk associated with low-dose chemical exposures. The models take into account only one part of the complex process and have no accepted basis in biology. The key decision points, qualitative factors, and quantitative considerations are identified and discussed and a risk assessment framework that incorporates these inputs is presented. A model that incorporates the tumorigenic dose-response information as well as the qualitative and quantitative biological factors that affect the estimate of risk is proposed. It is concluded that much work needs to be done before a completely quantitative approach to risk assessment is to be useful; in particular pharmacokinetic modeling should be pursued more aggressively.

INTRODUCTION

Society has made extensive use of chemicals to improve our standard of living. This use has resulted in a need to determine the risks associated with exposure to man-made and natural chemicals. Much discussion and disagreement on the subject of carcinogenic risk assessment and management have focused on the use of mathematical models for estimating potential risks. The purpose of this paper is to evaluate the current state-of-the-art of quantitative risk assessment, identify strengths and weaknesses of current methodologies, and provide an alternative procedure to stimulate discussion on this important subject. The alternative method is presented in the form of a simplified model that incorporates the quantitative dose-response information as well as other quantitative and qualitative biological data.

It is important to recognize that risk assessment encompasses three key areas: hazard identification, hazard evaluation, and risk evaluation (Figure 1, AIHC 1981). After a potential hazard has been identified, a hazard evaluation is made to determine from experimental data what kind of human responses could be expected under laboratory exposure conditions. The output of the hazard evaluation includes an estimate of the potential risk to humans. Since a hazard is not a risk to humans until there is exposure, a risk evaluation is made to determine whether and to what extent the potential hazard of a substance is realized in the real world. Risk evaluation

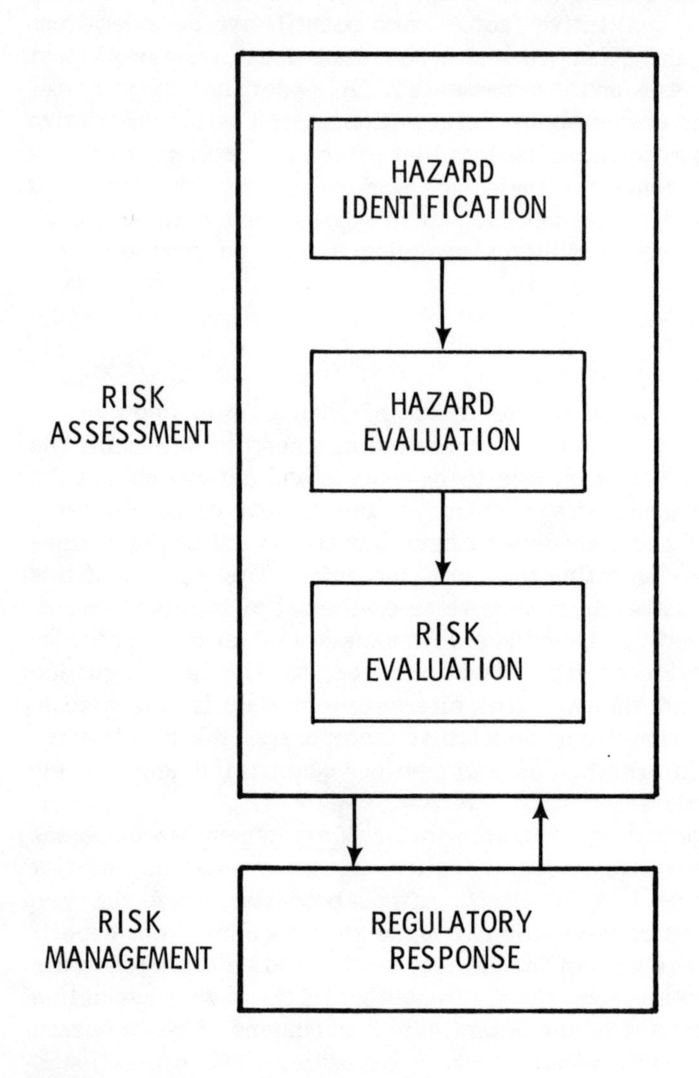

Figure 1. The four major steps in the process of risk assessment and management. AIHC uses the term Regulatory Response for the risk management step.

combines the results of the hazardous evaluation with information about the route and extent of exposure and number of persons exposed to estimate the extent of the risk to society.

Mathematical models have been widely used in the hazard evaluation step of risk assessment to estimate potential risk to humans from exposure to chemicals. It is, unfortunately, often assumed that risk assessment relies totally on mathematical models. This impression is partially due to the considerable discussion and controversy surrounding the use of mathematical approaches, as well as the fact that models have been over emphasized to the point of the exclusion of many important qualitative and quantitative factors necessary for the rational evaluation of risk. Quantification of potential risk and numerical determination of permissible exposure limits are often desirable but, as we will discuss, risk assessment must also incorporate a number of qualitative factors.

Risk management is the decision process that evaluates the benefits of the chemical versus the risks, costs and alternatives. It includes the regulatory response made to assessed risk (Figure 1). The result is the determination of how a given chemical can be safely manufactured, transported, used and disposed of with no significant impact on human health and the environment.

By its very nature, risk assessment is mainly a scientific activity while risk management is principally a political activity (Figure 1). The focus of this paper is on risk assessment, for we believe that sound risk management decisions will only result if a sound scientific analysis of the problem and data is made. It is important to keep these two activities separate because to do otherwise creates confusion and reduces the effectiveness of risk assessment and management.

Risk Assessment Process

Carcinogenic risk assessment is a complicated process that includes an analysis of the following types of studies in order to evaluate hazards and estimate the associated risks to humans (Figure 2).

- Mutagenicity
- Acute studies in animals
- Subchronic studies in animals

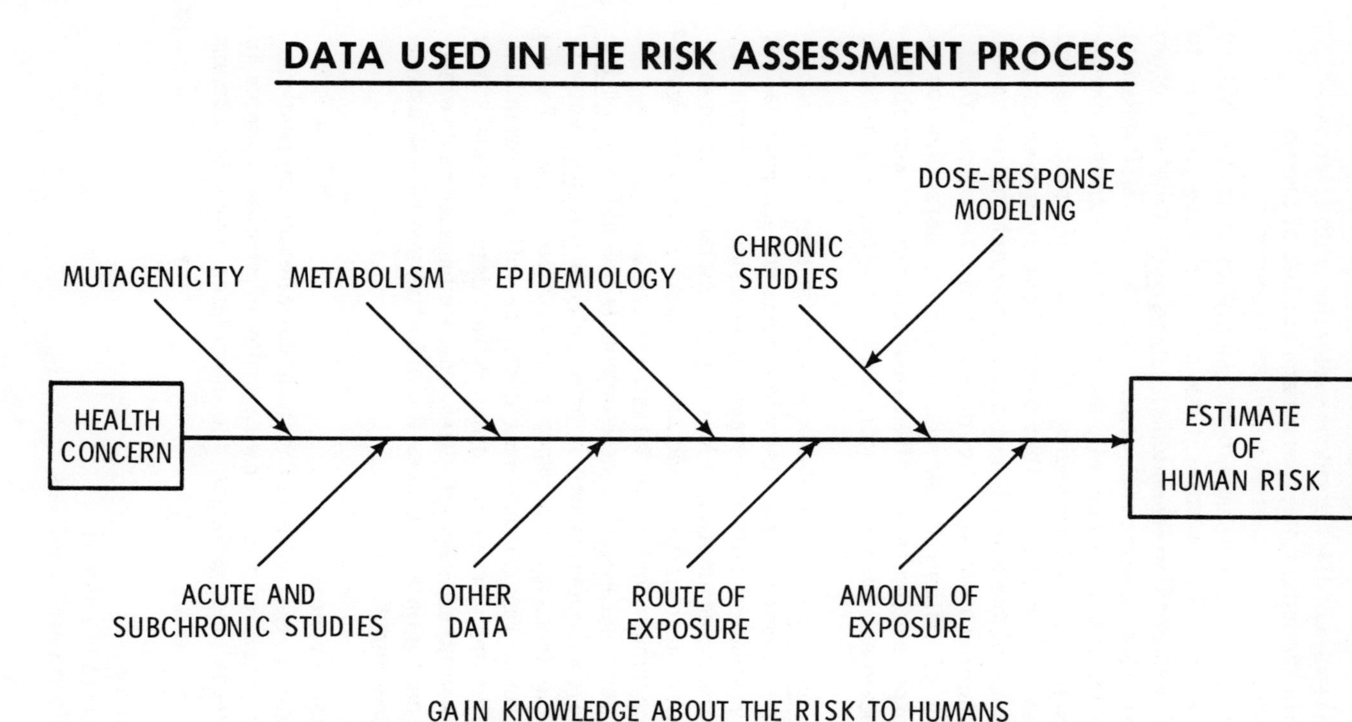

Figure 2. Data used in the risk assessment process. Dose-response models are used only in the analysis of the tumor incidence data collected in the chronic studies in animals.

- Metabolism
- Chronic studies in animals
- Epidemiology
- Route and amount of exposure

When these different types of data are organized, analyzed, and collectively interpreted, an understanding of the toxic potential of the chemical appears. Only after such an exercise has been conducted, do we have the scientific knowledge to decide how to handle a chemical. At present, formalized statistical models are commonly used only for chronic carcinogenic studies in animals. An overall model that incorporates data from the other types of studies has not been developed and likely will not be available in the near future.

The interest in evaluating the potential hazard of human exposures of many orders of magnitude below the experimental data base has led to the use of mathematical extrapolation models for predicting potential risk due to low-dose exposure. It was hoped that these models could characterize the low end of the dose-response curve for risk in laboratory animals and that the estimated dose-response would be useful for estimating human risk. Risk assessment by necessity, however, involves a number of qualitative decision points that are a very important part of the process and must not be overshadowed by the numerical aspects of risk assessment.

For instance, some compounds produce tumors early in life, at multiple sites and at doses with no other apparent toxicological response. In contrast, other chemicals produce a tumorigenic response only after overwhelming doses, very late in life, and often only at a single site where associated nontumorigenic toxicity is evident. These qualitative differences are important in assessing the potential for human hazard but they are not considered in the currently used mathematical models.

Some Views of the Current State-of-the-Art

There has been considerable discussion on the use of models in risk assessment and the biological basis for the models. Critics of a rigid modeling approach emphasize that results of chronic studies represent only part of the data base and that the biological processes are extremely complex. Squire (1981b) points out that

> "no models can actually be based on the biological events, since these are not known for any carcinogens."

This disparity between biological reality and the relevance of models has led to disagreement on the usefulness of the mathematical approach. It is generally agreed that a model must be based on fundamental biological mechanisms in order for its predictions outside the range of the observed data (i.e., extrapolations) to be of any value.

Carter (1979) reports that

> "Disagreement exists within the government itself over 'cancer policy' and especially over whether the science of quantifying cancer risks is far enough advanced for it to be safely used by regulatory agencies in setting standards for human exposure to carcinogens."

The National Academy of Sciences Committee that reviewed the pesticides regulation process (NAS 1980) studied the use of statistical models for describing results from limited bioassay data and reported

> "The Committee scrutinized carefully the methods used by OPP and EPA's Carcinogen Assessment Group. As a result of this scrutiny it came to the conclusion, very much in line with the position of the Director of the NCI (Carter 1979), that our present understanding of the mechanisms of cancer development does not permit us to draw reliable numerical inferences from the kind of laboratory data normally available about the effects of pesticides and other compounds on the development of cancers in humans."

Munro and Krewski (1981) commented in their review of risk assessment and regulatory decision making that

> "The use of point estimates of risk as a major decision criterion in the regulatory control of carcinogens would permit an assessment of the relative risks due to various compounds. Although this perhaps represents the ultimate

> application of mathematical modeling techniques, knowledge deficiencies in the science base presently preclude the possibility of realizing the full potential of the available procedures at this time."

Wahrendorf (1979) in an editorial on the estimation of safe dose levels concludes

> "The problem of safe dose estimation will definitely not be solved by the application of different mathematical dose-response models and the relevant statistical techniques of extrapolation alone. One should keep in mind that the isolated investigation of a noxa in a relatively homogeneous population of experimental animals, where the conditions are kept constant, is far removed from the situation of man in his environment."

These are only a few of the several instances in which scientists have expressed reservations concerning the use of potential risks calculated by statistical models. This does not mean that models should not be used. There are many cases in which it is necessary to arrive at permissible exposure limits and these estimates must be derived from existing quantitative data. The derivation of quantitative estimates from a quantitative data base implies the existence of a model, regardless of whether it is a simplistic model such as a safety factor or a more sophisticated approach. Modeling in one form or another will be used. It is important, however, to recognize the limitations and assumptions involved, and not to bury biological judgments in mathematical sophistication. A major concern is that no amount of disclaimer by the risk assessor will necessarily keep the risk manager from interpreting risk estimates at face value.

The strengths and weaknesses of currently available methodology do not seem to be well understood. The intent of this document is to summarize the state-of-the-art, bring together information from toxicology and statistics, and provide direction for future work. Box (1979) has emphasized that "all models are wrong but some are useful." All models are wrong because we never know the true state of nature. The relevant concern, therefore, is whether the model is illuminating and useful. We encourage our readers to keep this perspective in mind as they evaluate the use of models in risk assessment.

MODELS USED IN RISK ASSESSMENT

The models commonly used in risk assessment are based on either mechanistic or tolerance distribution assumptions.

Mechanistic Models	Tolerance Distribution Models
One-hit	Probit
Multistage	Logit
Multi-hit	Weibull

The use and assumptions of these models have been discussed in detail elsewhere (Bruce, 1980; Brown, 1978; Food Safety Council, 1980; Krewski and Brown, 1981) and will be reviewed only briefly here. Pharmacokinetic modeling, which is also a mechanistic approach, is discussed separately, elsewhere in this paper.

We should keep in mind that, while these models frequently fit the observed data equally well, they can predict very different potential risks at low doses. Krewski and Van Ryzin (1981) have shown mathematically, and by example, that the models tend to have the following relationship to each other at low doses:

Model	Estimate of Potential Risk
One-hit	Highest
Linear	
Multistage	
Weibull	
Multi-hit	
Logit	
Probit	Lowest

The statistical approach to selecting the best model would be to design the bioassay to discriminate among these models and identify the most useful candidate model(s). Unfortunately, as Crump

(1982) points out, it may not be possible to design such experiments. Clearly, the models disagree most in the low-dose region where the data contain no information (i.e., no excess tumors are observed). A brief description of some of the commonly used models is given in the following sections.

Mechanistic Models

One class of mechanistic models is derived from assumptions about the age specific tumorigenicity rate which is defined as the proportion of people in any specific age group (e.g., 46-50 years) developing cancer. It is also often referred to as the hazard function. If the age specific tumor rate (r) is a function of dose and age.

$$r = f(\text{dose,age})$$

and if the dose and age components can be separated into two mathematical functions so that

$$f(\text{dose,age}) = g(\text{dose})\ h(\text{age}),$$

then the cumulative lifetime risk for a given dose is found by integrating over age. The resulting class of models is

$$P = 1 - \exp(-g(\text{dose})),$$

where P is the probability of tumor and g(dose) is a mathematical function of dose.

One-Hit Model

This family of distributions includes the one-hit model for which

$$g(\text{dose}) = b_1(\text{dose})$$

and b_1 is a parameter to be estimated from data. This functional form assumes that for any given age, the probability of a tumor is directly proportional to the amount of exposure. This can result from the assumption that only one critical molecular event between a target site and the proximate carcinogen is sufficient to result in a tumor, and the probability of such an interaction is directly proportional to the nominal concentration of the carcinogen. This mechanism is, however, not the only one that is consistent with this equation.

The one-hit model and variations on it utilizing upper statistical limits (Gaylor and Kodell, 1980) represent a highly conservative approach to the extrapolation problem (Hoel, 1981). For example, a linear extrapolation of the Chemical Industries Institute for Toxicology formaldehyde study predicted that an average lifetime dose less than 0.66×10^{3} ppm was needed to keep the lifetime potential risk of tumor less than 10^{6} (Gibson, 1982). Such an estimate has little credibility as an estimate of the risk to humans when viewed in light of some 100 years of experience with human exposures to formaldehyde which generally are less than 0.1 ppm but have often been in the 0.1 to 5 ppm range (Todhunter, 1982) with no apparent increased carcinogenic risk.

The one-hit model assumes a dose-response that is approximately linear at low doses and concave downward over the entire dose range. The model, as commonly used, ignores the toxicological reality of non-linear dose-response mechanisms, saturation kinetics, no effect levels or thresholds of a real or practical nature.[2/] The one-hit model is, however, relatively insensitive to minor fluctuations in the data. It should be noted that linear extrapolation from the lowest observed effect level to a potential risk of 10^{6} (one in a million) generally results in a safety factor of approximately 100,000. Assuming an observed response level of 10 percent which is about the lowest detectable effect level in a standard bioassay, the safety factor = response/desired potential risk = $0.10/10^{6}$ = 100,000. This indicates that the dose corresponding to a potential risk of 10^{6} is approximately 1/100,000 of the lowest practically observable effect level.

The dose-response curve calculated from the one-hit model is essentially independent of the shape of the observed dose-response curve, gives very little weight to no-observed-effect-levels (NOEL), and may not be predictive of potential risk at low levels. When good epidemiology data are available for comparison, it has in some cases been found that the one-hit model is not compatible with the human experience (Gehring et al., 1979; Ramsey et al., 1979; Reitz et al.,

---

2/ We define a practical threshold as a dose so low that the probability of any tumors in the lifetime of a given population is negligible.

1978). Carlborg (1981a) concluded that the one-hit model did not give an adequate fit to most of the 31 data sets he analyzed. The one hit-model also does not fit the data obtained in the ED_{01} Study conducted by National Center for Toxicological Research (Carlborg, 1981b). This study used over 24,000 mice and was designed to define the shape of the dose-response curve at the low end, near one percent.

With appropriate species conversion, the one-hit model does, however, estimate an upper limit on the potential risk and may be useful in situations where an upper bound is of interest. For example, if the potential risk calculated by the one-hit model is not unacceptable, then there would be less need to consider other models. On the other hand, if permissible exposures predicted by the one-hit model are unrealistically low, which is often the case, then further risk analyses would have to be made to confirm or refute the one-hit model results. In all cases we must keep in mind that potential risks predicted by the one-hit model may be several orders of magnitude more than that of the true potential risk (factor of 10 = one order of magnitude).

Multistage Model

The multistage model assumes

$$g(\text{dose}) = (a_1 + b_1\text{dose})(a_2+b_2\text{dose}) \ldots (a_n + b_n\text{dose})$$

$$= c_0 + c_1\text{dose} + c_2\text{dose}^2 + \ldots + c_n\text{dose}^n$$

where a_1, b_1, $c_1 \geq 0$ are parameters that vary from chemical to chemical. The biological justification for this model is that cancer is believed to be a multistage process that can be approximated by a series of multiplicative linear functions ($a_1 + b_1$dose). This model is likely to be conceptually useful in some cases. In a theoretical sense, for instance, the concentration of the proximate carcinogen at the target site can be modeled by a series of kinetic reactions that are usually assumed to be linear at low doses but may be saturable (nonlinear) at high doses. Concentration kinetics may be linear at low doses. This does not necessarily imply a proportional response because at some concentration the existence of defense and repair systems is likely to modulate the response.

In practice, however, this results in fitting a polynomial model to the dose-response curve. The function generally fits well in the experimental dose range but has very limited applicability to the estimation of potential risk at low doses. The limitations arise first because the model cannot reflect changes in kinetics, metabolism and mechanisms at low doses; and second because low dose estimates are highly sensitive to a change of even a few observed tumors at the lowest experimental dose.

A logical statistical approach to account for the random variation in tumor frequencies is to express the results in terms of best estimates and measures of uncertainty. The following model provides an upper confidence limit on the potential risk but does not give a best estimate.

"Linearized" Multistage Model

The "linearized" multistage model proposed by Crump (1981) and utilized by many of the regulatory agencies replaces the linear term (b_1) of the polynomial function, g(dose), by its upper 95 percent confidence limit to reflect biological variability in the observed tumor frequencies. The dose-response predicted by this model is approximately linear at low doses (the d^2, d^3, etc., terms are essentially zero at low doses) resulting in estimates of potential risk that are almost identical to those of the one-hit model. Even for extremely nonlinear data (e.g., nitrilotriactic acid, Food Safety Council Report, 1980) estimated doses corresponding to potential risk levels of 10^{-6} differ only by less than a factor of 6 from the estimate of a one-hit model extrapolation. Thus, for almost all applications there is no appreciable difference between the linear model and the linearized multistage model.

This modeling approach relies totally on the upper confidence level for b_1 and ignores the best estimate of b_1 as well as the lower confidence limit. The result is that the model can produce a very high estimate of potential risk even when the dose-response function is very steep and when the exposure levels are well below no-observed-effect-levels.

The model has the potential to be applied even when the total dose-response is not statistically significant. We believe, however, that this can be very misleading and that low-dose extrapolation models should be used only when a statistically significant dose-response has been established.

Multi-Hit Model

The multi-hit model has been discussed in some detail in the Food Safety Council Report (1980). One derivation of this model follows from the assumption that K "hits" or molecular interactions are necessary to induce the formation of a tumor and the distribution of these molecular events over time follows a Poisson process. In practice the model appears to fit some data sets reasonably well and to give low-dose predictions that are similar to the other models. There are cases, however, in which the predicted values are inconsistent with the predictions of other models by many orders of magnitude. For instance, the virtually safe dose as predicted by the multi-hit model appears to be too high for nitrilotriacetic acid and far too low for vinyl chloride (Food Safety Council Report, 1980).

Tolerance Distribution Models

Another approach to the modeling problem is to assume that each member of the population will develop a tumor if exposure to the carcinogen exceeds a critical level. This threshold level varies from individual to individual and has been modeled by various tolerance distributions.

Log-Probit Model

The log-probit model assumes that the individual tolerances follow a lognormal distribution. Specific steps in the complex chain of events that lead to carcinogenesis are likely to have lognormal distributions. For example, it is reasonable to assume that the distribution of a population of kinetic rate constants for detoxification, metabolism, elimination, as well as the distribution of immunosuppression surveillance capacity and DNA repair capacity can be adequately approximated by normal or lognormal distributions.

Tolerance distribution models have been found to adequately model many types of biological dose-response data, but it is an overly simplistic expectation to represent the entire carcinogenic process by one tolerance distribution. A tolerance distribution model may give a good description of the observed data but from a mechanistic point of view there is no reason to expect extrapolation to be valid. The probit model extrapolation has, however, fit well in some instances (Gehring et al., 1979).

Logit and Weibull Models

Other tolerance distributions which have been used to model carcinogenicity dose-response data include the logit and Weibull models (Doll, 1971; Carlborg, 1981a). The multi-hit model discussed earlier can also be viewed as a member of this class of models that uses the gamma function to model the tolerance distribution. For this reason it is often called the gamma-multi-hit model.

The log-probit, logit, Weibull, and gamma distributions all have potentially similar shapes between tumor frequencies of 2 percent to 98 percent; hence, it is not surprising that these models often give essentially identical fits to the observed data, but again, the models differ widely at low doses.

OTHER USEFUL MODELS

No-Effect-Level Model

The observation that many biological responses vary linearly with the logarithm of dose, and that practical thresholds exist, can be represented by the following model:

$$\text{Response} = B_1 \qquad \text{if dose} < d^*$$

$$\text{Response} = B_1 + B_2 \log(\text{dose}/d^*) \qquad \text{if dose} \geq d^*$$

This model incorporates a parameter d^* that represents a threshold below which no dose-response occurs. In this model B_1 is the constant response level at doses less than d^* and B_2 is the slope of the logdose-response curve at doses $\geq d^*$. It has been empirically found that many quantitative toxicological end points can be adequately described by the no-effect-level model. This model may, therefore, be useful for establishing thresholds for end points related to the carcinogenic process in situations where information other than the simple presence or absence of a tumor is available. Both the model and predicted threshold are of value when carcinogenicity is a secondary event. The use of this model in estimating no-observed-effect-levels is discussed further in the section entitled "Alternative Analyses of Chronic Studies."

Pharmacokinetic Models

Pharmacokinetic models have often been used to predict the concentration of the parent compound and metabolites in the blood and at reactive sites, if identifiable. Cornfield (1977), Gehring and Balu (1977), and Anderson et al. (1980) have extended this concept to include rates for macromolecular events (e.g., DNA damage and repair) involved in the carcinogenic process. The addition of statistical distributions for the rate parameters and a stochastic component representing the probabilistic nature of molecular events and selection processes may represent a useful conceptual framework for describing the tumorigenic mechanism of many chemicals. Pharmacokinetic data are presently useful only in specific parts of the risk assessment process. A more complete understanding of mechanism of chemically induced carcinogenesis would allow a more complete utilization of pharmacokinetic data. Pharmacokinetic comparisons between animals and humans are presently most useful for making species conversions and understanding the qualitative and quantitative species differences. The modeling of blood concentrations and metabolite concentrations identifies the existence of saturated pathways and adds to the understanding of the mechanism of toxicity in many cases.

In the future as more is understood about the mechanisms of carcinogenesis, formalized quantitative approaches incorporating pharmacokinetic data will likely become more useful in risk assessment.

LIMITATIONS OF THE MODELS

The models described above, with the exception of pharmacokinetic approaches, are generally an oversimplification of a complex system and apply only to the chronic animal toxicity studies that are but one input to the risk assessment process. The models have little biological relevance, have been shown to provide poor extrapolation estimates and, with a few exceptions, have not been validated either in animals or with respect to the human experience. Many of these models can be used, however, to summarize the dose-response curve within the range of observable responses and pharmacokinetic models are of use for understanding and predicting specific parts of the overall process.

Two major problems with the use of formalized modeling approaches are, (1) the inability of models to incorporate much of the qualitative information that must be used to arrive at logical decisions, and (2) the statistical problems involved in extrapolating quantal data.

Incorporating Qualitative Information

The importance of the qualitative decisions is evidenced in the process by which both governmental agencies and industry determine how to handle a potential carcinogen. It is initially a two-step procedure in which a qualitative decision is made as to whether the compound is an animal and/or potential human carcinogen. If the answer to either of these questions is yes, then the risks to humans are estimated and a decision is made as to how to handle the compound. If a complete risk estimation model existed, there would be no need for the first step. The model would be used to compute the risks directly and noncarcinogens would be assigned very small or zero risks. Unfortunately, our current state of knowledge does not permit us to take this approach.

Type of Tumor is Important

There are many specific qualitative decision points that impact the quantitative estimation of potential risk. These qualitative data include the type of tumor observed in the experimental animal. The B6C3F1 mouse, for example, is highly sensitive to hepatic tumors and the majority of male Fischer 344 rats develop testicular tumors independent of any chemical exposure. Chemicals that increase the incidence of these tumors but do not show any other tumorigenic activity should be considered as possible promoters of tumorigenesis. Such chemicals are considered to be much less likely to cause human cancer than a potential carcinogen that produces a spectrum of histogenically different tumors including those with low spontaneous rates. The mechanisms of action for some promoters is entirely different from that of complete carcinogens and certain steps in the assessment of risk should reflect the different mechanisms.

Examples of chemicals that only increase the spontaneous liver tumor rate for B6C3F1 mice after chronic high exposures, and are apparently not carcinogenic in rats, and do not appear to be mutagenic or genotoxic include the chlorinated solvents, perchloroethylene, and trichloroethylene. These substances do not appear to present a significant carcinogenic risk to humans at present exposure levels.

Relevance of Tumor Type

A similar type of qualitative information that impacts estimates of potential risk is the relevance of the observed tumors in animals to potential human hazard. Examples of tumors that have limited predictability to humans from the animal model include zymbal gland carcinomas and tumors of the nasal turbinate. An excellent example is formaldehyde which produces tumors apparently only in conjunction with chronic irritation of the nasal passageways. The rat, as an obligatory nasal breather, is sensitive to chronic irritation of the nasal mucosa and develops tumors in this area after chronic exposure to a number of chemicals. While it is possible, of course, that excess human exposure would result in respiratory tumors, models used to estimate potential risk must be capable of appropriately incorporating this qualitative observation. Mathematical models currently in use do not have this capability.

Genotoxic and Nongenotoxic Mechanisms

Probably the most important qualitative difference between animal carcinogens is the distinction between carcinogens that operate primarily through direct genotoxic mechanisms and those that produce tumors through mechanisms other than direct interaction with DNA (i.e. epigenetic, nongenotoxic, or nongenetic mechanisms). The subject of tumorigenic mechanisms has received a great deal of discussion (Weisburger and Williams, 1980; Stott et al., 1981). Briefly stated, some chemicals that are carcinogenic in animals also interact directly with DNA as indicated by results of short-term in vitro tests, or by direct measurement in vivo DNA alkylation and repair rates. Other chemicals that are tumorigenic in animals show virtually no activity in the genotoxicity tests and show no propensity to bind or interact with DNA. Research on some of these chemicals (e.g., saccharin, chloroform, trichloroethylene, and perchloroethylene) has shown that there is little to no direct interaction with DNA. On the other hand, in vivo tests indicate a dose-dependent acceleration in the rate of DNA synthesis at doses that correlate well with tumorigenicity and demonstrates an apparent tumorigenic mechanism and threshold (Stott et al., 1981; Reitz et al., 1980; Schumann et al., 1980).

In some of these cases the likely mechanism of action is cytotoxicity resulting in cellular regeneration accompanied by an increased rate of DNA synthesis. DNA is constantly undergoing a low background rate of damage and subsequent repair. If this background rate is increased (e.g., when cellular damage necessitates an increase in the rate of DNA synthesis), the increased demands on the repair surveillance systems may lead to an increased probability of faulty DNA repair or the possibility of replication before repair is completed. This phenomenon can be demonstrated by the production of skin tumors following burns and the repeated freezing of skin with dry ice, (Bernblum, 1929; Laroye, 1974) and liver tumors following partial hepatectomy.

Cytotoxicity and direct cellular damage leading to tumorigenesis is only one nongenotoxic mechanism for carcinogenesis. Weisburger and Williams (1980) point out that other nongenotoxic mechanisms and examples include solid state carcinogens (polymers, asbestos), hormonal imbalance carcinogens (estradiol, DES), immuno suppressors (azothioprine), and promoters (phorbol esters, saccharin).

Use of Mechanistic Information

The subject of tumorigenic mechanisms is extremely complex. For example, when classifying chemicals with respect to carcinogenic mechanisms, we should keep in mind that genotoxicity is a continuous spectrum rather than a dichotomous classification. Some chemicals clearly have a direct genotoxic component while others are at or near the nongenotoxic end of the scale. When doing risk assessment, it is important to identify those compounds that apparently operate through mechanisms other than direct genotoxicity. The importance of the distinction between genotoxic and nongenotoxic mechanisms is that, according to current theory, all that is needed for directly genotoxic chemicals to initiate a tumor is a single molecular event; thus, thresholds may not exist. We note, however, that the existence of protective mechanisms, such as DNA repair systems that are saturable, would imply the existence of a practical threshold.

On the other hand, for nongenotoxic mechanisms, more than a single molecular interaction is necessary to produce a tumor. The consensus of scientific opinion is that, for some of these mechanisms, as with other toxicological phenomena, either an absolute threshold exists or the dose-response curve is so flat as to be indistinguishable from zero slope at low doses (i.e., a practical threshold exists). In some cases, the existence of an observable precursor related to tumorigenicity (e.g., cellular toxicity, necrosis, and hyperplasia), or a change in the rate of DNA synthesis may provide a marker variable that can be used to predict the threshold below which the nongenotoxic mechanism ceases to pose a risk.

For instance, many chemicals increase the spontaneous rate of hepatic tumors in B6C3F1 mice. High doses of many of these compounds result in tumorigenicity that is often preceeded in dose and in time by microscopically observable histological alterations. For these chemicals the experimentally observable tumor frequency data are often augmented by earlier quantitative signs of clinical and subclinical liver toxicity that can be most useful in establishing better measures of no-observed-effect-levels. In those cases, a practical threshold exists as well as a mechanism for observing and quantifying the threshold.

These chemicals should be viewed as promoters of tumorigenesis in animals and should be evaluated as such in the traditional manner of other quantitative toxicological phenomena with demonstrated no-effect levels. Knowledge of species differences in metabolism and all the accompanying scientific judgments should be used in estimating the corresponding safe level for humans.

Extrapolation of Quantal Data

Another major limitation of the models is the large statistical uncertainty in extrapolating data orders of magnitude below the observable range, particularly for quantal responses. Two sources of variation that contribute to the uncertainty inherent in tumor frequencies are statistical (binomial) variability which implies a large degree of variation in tumor counts and experiment-to-experiment variability.

Tumor Identification Is Based On Subjective Judgment

We should also keep in mind that the variation observed in tumor frequency is a function of the uncertainty involved in classifying lesions into categories of "tumors" and "non-tumors"; such as the subjective judgments used in separating high grade hyperplasia of the liver from low grade carcinomas. It was emphasized in the various examinations of data in the ED_{01} Study conducted by the National Center for Toxicological Research that the distinction between carcinomas and "non-tumors" such as hyperplasia is a subjective judgment and may not be consistent from pathologist to pathologist (Squire, 1981a). It is also important to recognize that this judgment is often most difficult to make at low doses because a dose-response may exist in the severity of lesion as well as the frequency of lesions. At high doses malignancy is often clearly defined; however, at low doses the responses are often equivocal and may be classified as tumors or nontumors, depending on the opinion of the pathologist.

Quantitative Measures of Carcinogenesis Are Needed

A large source of statistical variability in the estimated potential risks is due to the inherent problem of extrapolating quantal responses. When information available on each animal is only the determination of presence or absence of a particular tumor, the standard error of the quantal estimate is large when compared to the standard error of quantitative measurements such as organ weights or clinical chemistries.

Recent theoretical work and simulation studies (Brown, 1978; Krewski, 1981) have shown that variation between models for the same data sets and variation in the estimates of potential risk produced by the same model as a result of random binomial variability can result in low-dose risk estimates varying by orders of magnitude. The problem of correctly specifying a model in the absence of biological mechanisms, combined with the inherent variability in low-dose extrapolation places severe limitation on the usefulness of a rigorous modeling approach. This is not to say that dose-response modeling should not be done; there are situations in which quantification is necessary and the information is useful for other than regulatory purposes. It is important, however, to understand the limitations of quantitative approaches so that computer generated estimates of potential risk are not given more credibility than they deserve.

Determination of the Dose to the Target Tissue is Critical

Another reason for the large uncertainty and inconsistency noted in the extrapolation process is the measurement of dose or exposure. The dose used in the modeling should be that which is seen by the target tissue rather than the nominal dose administered to the animal. This is especially important if the exposure route in the experiment is different from that of the human. If the nominal and effective doses are strictly proportional, then nominal dose is an appropriate surrogate. Often, however, the high doses used in a bioassay saturate normal detoxification and excretory mechanisms resulting in nonlinear relationships between nominal and effective doses. In this case, Michaelis-Menten kinetics are observed and may be accompanied by alternative metabolic and excretory pathways that can result in enhanced toxicity of the chemical (Gehring and Blau, 1977). The overall consequence of nonlinear pharmacokinetics is that toxicity may increase disproportionately with increasing dose. Thus, extrapolation from high doses at which detoxification is overwhelmed to low nonsaturating doses using models that do not account for nonlinear kinetics can greatly overestimate the potential risk.

The issue becomes even more complicated when one relates the animal studies to humans and tries to determine the appropriate target tissue in humans and the dose of the chemical seen by those tissues. Metabolic and pharmacokinetic data can help answer these questions, but unfortunately, almost all dose-response modeling reported to date is based on doses administered to the animal with no regard for kinetic data (Anderson et al., 1980).

STATISTICAL MODELING PROCEDURE IS INCOMPLETE

The statistical modeling procedure, as it is currently applied to the results of chronic studies, is incomplete because it does not account for differences in tumor site, adequacy and completeness of the study, sex, strain, or data regarding differences in species sensitivity. For example, a study of the effects of a chemical on five organs in male and female rats and mice in two laboratories would commonly result in $5 \times 2^3 = 40$ estimates of potential risk. Little critical thought has been given to summarizing this information into a single, or not more than a few, estimates; hence, a popular procedure has been to report the highest estimate. This sets a conservative upper bound on the true potential risk. There is a need, however, for more accurate estimates so that informed risk management decisions can be made.

Results of Negative Studies Should Not Be Ignored

It is often the case that for a particular chemical some studies show a positive tumorigenic response while others show no effect. These so-called "negative studies" may be due to the use of inadequate experimental designs, the result of small numbers of animals, or, as is often the case, may represent differences in species or strain sensitivities. As discussed in the next section, the results of negative studies can be combined with those of other studies to produce an overall estimate of potential risk that has higher precision than that of any of the individual studies. It is a fundamental principle of statistics that the average is usually a more precise estimate of an unknown true value than any subset of the measurements used to compute the average.

Using All the Data to Estimate Potential Risk

Quantal data on the frequency of observed tumors in groups of experimental animals provide only very limited statistical information. Thus, it is wasteful, as well as poor scientific practice, to use the study that gives the highest estimate of potential risk. This practice, along with ignoring negative studies and studies based on small sample sizes, has the potential to unnecessarily bias the results and will reduce the precision of the estimated potential risk. There are at least two ways to combine the results of chronic studies thereby increasing the precision of the risk estimate; 1) compute weighted averages of risk estimates and 2) use statistical models which allow for differences among data sets.

A generally accepted procedure for combining estimates from several studies is to compute a weighted average of the estimates. The weights commonly used are inversely proportional to the variances of the estimates. This approach uses all the data, incorporates the uncertainties associated with the data, and can be extended to include both within-study and between-study variance components (St. John et al., 1982).

Another approach to combining results from different sexes, studies, and strains is to incorporate parameters in the model to account for these differences (Carlborg, 1981a). As described by Carlborg, chi-square tests are available to test for the appropriateness of this procedure. The result is models that utilize all the available information and produce more precise estimates of the potential risk at low-dose exposures.

BEST ESTIMATES AND BOTH CONFIDENCE LIMITS SHOULD BE REPORTED

Best estimates as well as lower and upper confidence limits should be routinely reported. The use of only the upper confidence limit on potential risk as the appropriate measure of risk is a political decision. When this is done, it results in political concerns and judgments becoming part of the risk assessment process which is inherently a scientific activity. The size of the confidence limits is inversely proportional to the quality of the data used to make the estimate and directly proportional to the amount of extrapolation involved. This important information is lost if the confidence limits and best estimates are not routinely reported. The width of the confidence interval is one of the best measures risk assessors and risk managers have to evaluate the quality of the estimates of potential risks. It is important to distinguish between those situations in which the risk is precisely estimated and those in which it is not.

As noted earlier, the use of only the upper confidence limit for the linear term (b_1) in the multistage model can result in a nonzero estimate of potential risk for a noncarcinogen. Reporting of the best estimate and both the lower and upper confidence limits for b_1 would alert the risk manager to the fact that the dose-response relation for the chemical was not statistically significant.

Confidence limits have some limitations that may not be well recognized. The confidence limits are statistically correct only if the model used to compute the limits is an accurate representation of the underlying dose-response function. Confidence limits do not provide a measure of modeling error, i.e., the extent to which the model is correct. A measure of this uncertainty can be obtained, however, by fitting a number of different models including the calculation of their associated confidence limits (Hoel, 1981).

ALTERNATIVE ANALYSES OF CHRONIC STUDIES

It should also be recognized that there are ways of evaluating the results of chronic studies other than fitting rigorous statistical models and extrapolating orders of magnitude away from the region of the data. Some of the more useful ways are: estimation of a no-observed-effect-level (NOEL), comparison of chemical dose-response to that of other chemicals, and use of time-to-tumor information. Each of these approaches is discussed in the following paragraphs.

Estimation of No-Observed-Effect-Levels

No-observed-effect-levels (NOEL) have been used by toxicologists for years to evaluate the potential risk due to exposure to chemicals. The use of NOEL's is often criticized because they are dependent on the number of animals tested at each dose level and the specific dose levels used in the experiment. The dependence on the dose levels tested is eliminated by fitting a statistical model that includes the NOEL as a parameter. The uncertainty of the estimated-no-observed-effect-level (E-NOEL) directly reflects the number of animals tested at each dose level. The E-NOEL is interpreted as the dose level at which no effect would have been observed if that dose level were tested in the study and if the model is correct.

In mathematical terms the E-NOEL could be estimated using the model

$$P = B_1 \qquad \text{if dose} < \text{NOEL}$$
$$P = B_1 + F(\text{dose}) \qquad \text{if dose} \geq \text{NOEL}$$

where P is the probability of tumor, B_1 is the background, and F(dose) is the dose-response model used to describe the data. For example, in the case of a linear dose-response, the model would be

$$P = B_1 \qquad \text{if dose} < \text{NOEL}$$
$$P = B_1 + B_2(\text{dose-NOEL}) \qquad \text{if dose} \geq \text{NOEL}$$

where NOEL is an unknown parameter whose estimate and uncertainty is determined by fitting the model to the data and B_2 is the slope of the dose-response curve for dose $\geq$ NOEL. The log-dose-response model was illustrated in Section 2.3.

The dose-response data on bladder tumors in mice due to exposure to 2-acetylaminofluorene (2-AAF) is a good example where it is appropriate to consider the estimation of a NOEL (Figure 3, data obtained from Bruce, 1980). The model:

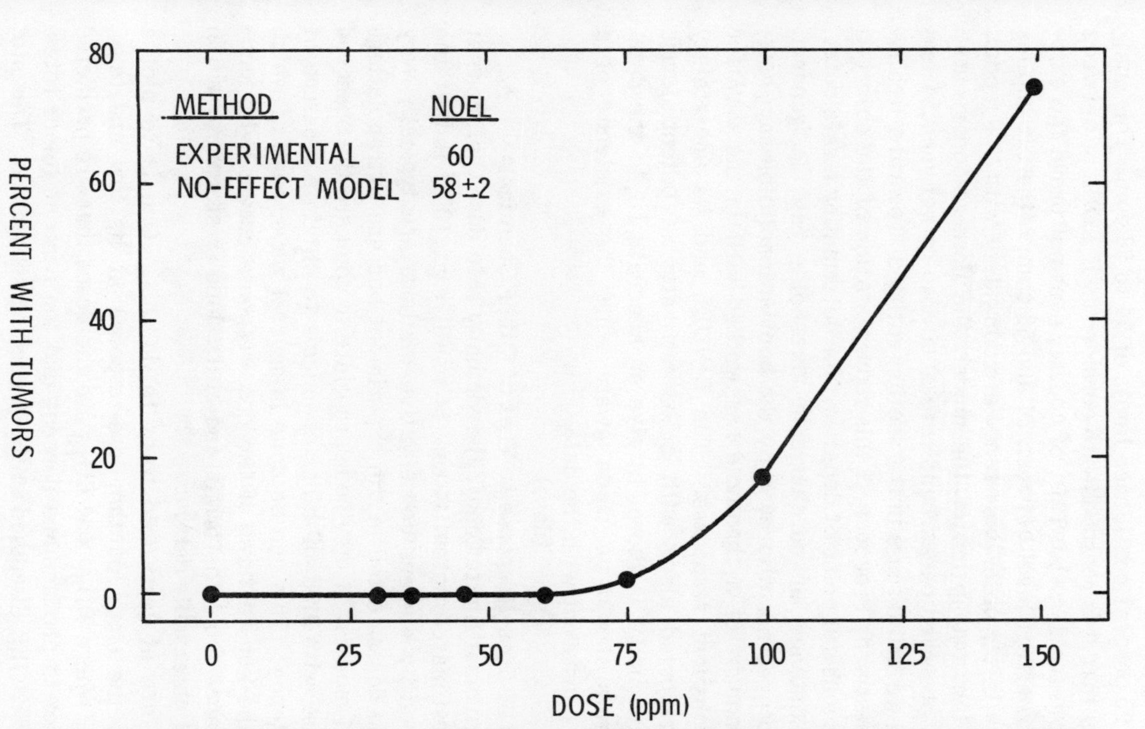

Figure 3. Dose-response of 2-AAF for bladder tumors in mice (data from Bruce, 1980). The no-observed-effect-level is 60 ppm. The estimated-no-observed-effect-level (E-NOEL) based on the no-effect-level model is 56 ppm with 95 percent confidence limits of ± 2 ppm (see sections entitled "Models Used in Risk Assessment" and "Alternative Analyses of Chronic Studies").

$$\ln(P/(1 - P)) = -5.53 \qquad \text{if dose} < 56$$
$$\ln(P/(1 - P)) = -5.53 + 6.75 \ln(\text{dose}/56) \qquad \text{if dose} \geq 56$$

was found to give an adequate fit to the data with an E-NOEL of 56 ppm and 95 percent confidence limits of 54 to 58 ppm. This analysis suggests that it is reasonable to consider an E-NOEL of between 54 and 58 ppm for 2-AAF. This, of course, does not prove the existence of a no-effect level between 54 and 58 ppm; rather, it implies that the data in the observable range are consistent with this model, and, under the assumption of the model, the dose-response curve reaches the observed background rate at a dose of between 54 and 58 ppm. As we will discuss in the section entitled "Toward a Realistic Risk Assessment Procedure," this representation of the observed data may be a useful point of departure for determining a safe dose.

The existence of carcinogenic thresholds can, in general, neither be proved nor disproved by the conventional bioassay; however, the concept of a threshold has worked well in the past for many toxicological responses. The E-NOEL and its uncertainty should be reported along with other estimates of potential risk whenever the model is shown to give an adequate fit to the data. Risk managers should be made aware when the existence of a threshold is consistent with the data.

Comparison of Dose-Response to That of Other Chemicals

The principal limitation of the virtually safe dose (i.e., dose at which the estimated potential risk is small, e.g., 10^{-6}) is that the VSD produced by current conservative methods are typically very far from the doses tested in the experiment and are extrapolations of empirical models. It may be appropriate to compare the observed dose-response with that of other chemicals to obtain a measure of relative potency. This can be done in several ways, each of which relates to the key features of the dose-response curve; 1) its location or distance from the origin and 2) its slope or steepness within the range of observable data.

There are at least two ways to do this. One is to plot a measure of the slope of the dose-response at its low end (e.g., ED_{10}/ED_{01}, where ED_{01} and ED_{10} are the doses that are predicted by the model to produce one percent and ten percent tumor rates) versus the E-NOEL discussed in the preceding subsection. This plot would be limited to those situations where it was possible to obtain an E-NOEL.

Another possibility is a plot of the location and slope coefficients in a fitted model. In the case of the Weibull model,

$$P = 1 - \mathrm{Exp}(A + B(\mathrm{dose})^{M})$$

one would plot B versus M. Carlborg (1981a) has shown that this model, with background response, slope, and location parameters A, B, and M, respectively, can describe the dose-response of a wide variety of chemicals. This plot and the one mentioned in the previous paragraph could be used to compare the dose-response of "unknown chemicals" to the dose-response of chemicals whose carcinogenic potential is better known and understood.

One of the key advantages of these two approaches is that they allow both the risk assessor and risk manager to visually evaluate the available data and relate it to data on other chemicals rather than make a decision based on a computer-generated estimate of potential risk that is typically orders of magnitude away from the data. It should be noted that these plots will be meaningful only if the dose of each chemical is expressed in the same units.

Time-To-Tumor Information Should Be Included When Available

Recent analyses (Sielken, 1981; Salsburg, 1981; Van Ryzin, 1981) of the Ed_{01} study have pointed out that time-to-tumor information affects the shape of the dose-response curve, thereby influencing the estimates of potential risk produced by the models. It is recognized that in some instances accurate time-to-tumor information can be difficult to get without performing serial sacrifice experiments which may increase study costs. It is clear, however, that the determination of time-to-tumor is important in carcinogenicity studies and this information should be included quantitatively or qualitatively whenever possible.

The use of time-to-tumor data can also provide alternative measures of potential risk such as the "late risk dose" which is the dose corresponding to an acceptably late increase in risk and "mean free dose" which is a measure of the percent of lifetime free from tumor (Sielken, 1981). These measures of potential risk are attractive because they are easy for people to understand. Further work is needed to better determine the utility of these approaches to risk modeling.

TOWARD A REALISTIC RISK ASSESSMENT PROCEDURE

The risk of developing cancer from chemical exposure cannot be estimated by a single number calculated by modeling the tumor incidence rates obtained in chronic studies. A number of different approaches is necessary. These will by necessity take into account several pieces of the risk assessment process (Figure 2) because an overall model is not available. The result will be a collection of qualitative and quantitative information that will be the basis for the final decision.

The following paragraphs first review proposals in the literature and then summarize the key decision points that are often not adequately considered in risk assessment. This discussion is followed by a quantitative risk assessment procedure that we present for consideration. The procedure incorporates the qualitative and quantitative data discussed in this paper and may be appropriate as a conceptual approach to risk assessment.

Key Aspects of Risk Assessment

Squire (1981b) includes the following types of data in his method for ranking animal carcinogens:

- Number of different species affected
- Number of different types of neoplasms
- Spontaneous tumor incidence in control groups
- Dose-response relationships
- Malignancy of induced neoplasms
- Genotoxicity

It is significant that, with the exception of dose-response relationships, these sources of information are not taken into account by any of the commonly used statistical models. Squire's ranking procedure considers only animal data. Human data should also be incorporated whenever they are available.

Weisburger and Williams (1981) also note the importance of understanding the mechanisms important to the carcinogenesis process and recommend an approach to carcinogen testing that is based heavily on the use of short-term tests _in vitro_ to determine genotoxicity. Clearly, risk assessment must take genotoxicity into account but, unfortunately, the statistical models have not been extended to incorporate this critical information.

This paper has identified several qualitative factors and quantitative considerations that are critical to the risk assessment process. These are discussed below with particular reference to the type of data needed and the decision points involved. It is useful to collect these qualitative aspects into a "decision tree" framework such as in Figure 4. A decision tree requires clear-cut yes/no decisions that may be inappropriate for some of the factors being considered (e.g., genotoxicity) but it is a useful concept for making some of the decisions required in risk assessment.

Qualitative Factors

Animal carcinogenicity. The first decision is whether the compound is an animal carcinogen through any mechanism. This involves examining all the relevant animal studies and evaluating whether the data base indicates a treatment-related positive response in any species. The critical decision in this step is to avoid false positives and false negatives in making a valid judgment concerning the likelihood of a repeatable treatment-related response.

Potential for human carcinogenicity. This is an important step in the evaluation process that is often skipped. Many compounds have been shown to cause cancer in laboratory animals under specific conditions, but there are few proven human carcinogens possibly due to the lack of sensitivity of many epidemiology studies or the difference in human response at low exposures. Of course, the default assumption must be that humans will be similar to the appropriate animal model, but this does not imply an automatic decision. This step involves the qualitative and quantitative evaluation of comparative organ toxicity, pharmacokinetics, (absorption, distribution, metabolism, and excretion over time), similarity of routes of exposure and the consideration of results from short-term tests. An important qualitative consideration is the judgment as to the relevance of route of exposure. Risk assessment should not use the results of bioassays in which the route of exposure to the sensitive organ is not comparable to potential human exposure or when there is not sufficient metabolism data to compare routes of exposure. Epidemiological data are the most relevant evidence for judging the likelihood that positive results observed in laboratory animals are indicative of potential risk to humans. It must be recognized that, while interpretation of these data is often not simple, they often do provide valuable information.

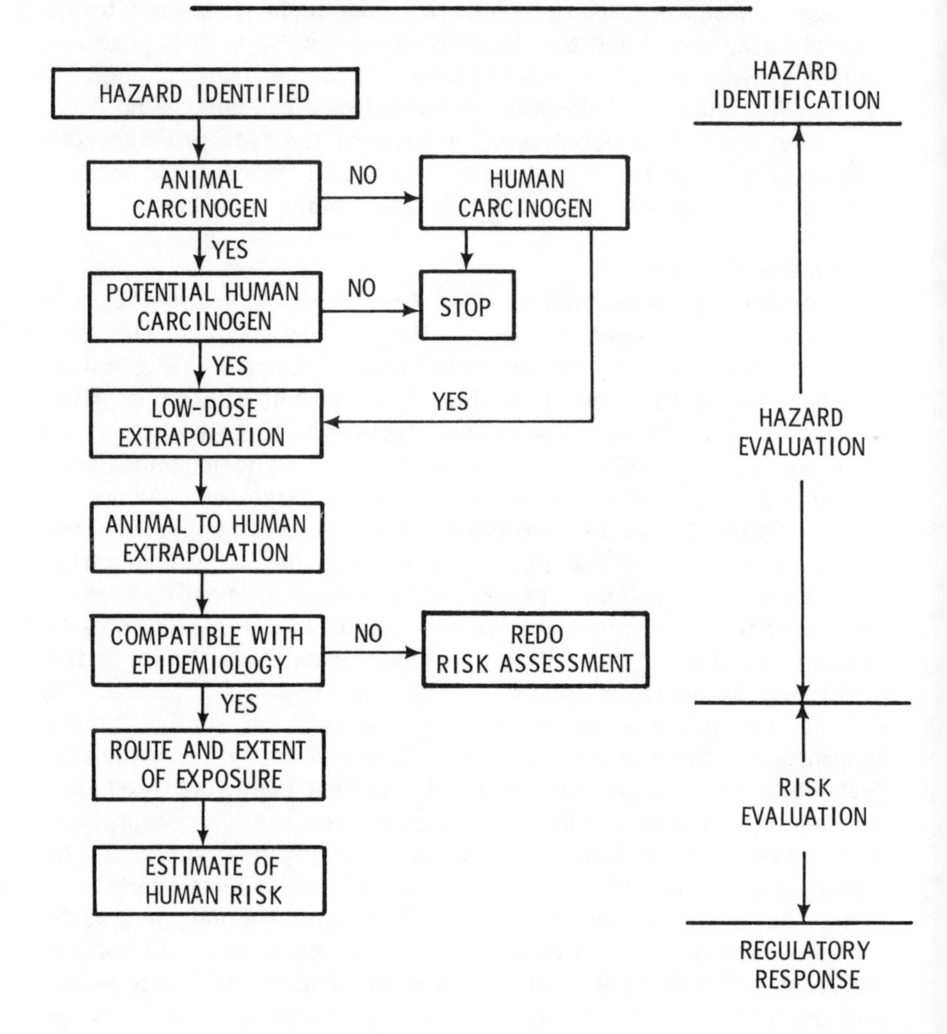

Figure 4. Key aspects of risk assessment identifying major decision points and showing their relation to AIHC's four step risk assessment and management process.

Genotoxicity. The massive doses used in laboratory carcinogenicity studies increase the probability for the expression of cancer through nongenotoxic mechanisms such as that of recurrent cellular toxicity. If sufficient data exist to demonstrate that a particular chemical does not interact with DNA in vivo or in short-term tests (e.g., Ames test, unscheduled DNA synthesis test, mammalian cell mutagenesis, etc.), and that the likely mechanism of action is a simple expression of localized toxicity, then the risk assessment should proceed on the basis of the toxicity of the chemical at the site of action rather than the assumption that the chemical is a direct acting systemic carcinogen. In this case the risk assessment should involve models relevant to toxicity, including log-linear and log-probit models, and safety factors.

Nitrilotriacetic acid is one example of a chemical with a relatively large data base that strongly supports the concept of cancer caused by a nongenotoxic mechanism and the existence of a threshold dose. The tumorigenic dose-response is only apparent at high doses and drops off rapidly to a no-observed-effect-level. All the data indicate little or no direct genotoxic acitvity (Anderson et al., 1982).

An independent multidisciplinary science panel would be especially useful at this point in the decision-making process since the consideration of qualitative factors involves subjective evaluation of data from diverse areas.

Quantitative Considerations

Animal to human extrapolation. Regardless of the method of extrapolating from effects at high doses to effects at low doses, a conversion must be made from animal to human. This must be done using an estimate that incorporates all the appropriate data, rather than using conversion factors that are not based on data. In the rare instance where comparative metabolism or pharmacokinetics data are available, these data must be used to make the animal to human extrapolation.

The general regulatory assumption has been that humans are more sensitive than laboratory animals on a mg/kg basis. This is based upon basic metabolic rates and is quantified by the so-called surface area rule. It is possible that this argument may be justifiable for some direct acting chemicals that require no metabolic activation. However, in the case of chemicals requiring metabolic activation, it can be argued that the correction factor should be just the reverse (Reitz et al., 1978). The peak concentration and total body burden of metabolic products will be higher for the laboratory animals because they metabolize the chemical quickly whereas humans will, in general, form lower concentrations because of the competing effects of elimination and metabolism. In general, it may not be appropriate to use the surface area correction factor in reverse for those chemicals requiring metabolic activation but it is implausible to assume that humans are always more sensitive than laboratory animals. This is particularly true when multiple species or strains have been tested and the only tumorigenic response is hepatic tumors in the highly sensitive B6C3F1 mouse. The point is that when interspecies conversion data are available, they must be used rather than relying on a simplistic conversion formula regardless of the data.

Epidemiology. After the animal-to-human extrapolation has been completed, it should be validated against epidemiology data when such data exist. If it can be shown that the model overpredicts or underpredicts the epidemiology experience, then the analysis must be re-done, using the human data as one input.

We have emphasized that quantitative risk assessment, in one form or another, is necessary but that the strict adherence to one specific dose-response model or class of models has failed to adequately characterize potential risk for a number of reasons. These reasons include the inability of present models to incorporate important qualitative data, the inherent problems of extrapolating quantal data far outside the observable range, and the lack of biological foundations for the models. The following approach to risk assessment is an attempt to formalize the considerations made in this section and to use information from the fields of toxicology and statistics to arrive at a logical decision making framework. The framework utilizes dose-response data, but is not greatly dependent on the choice of models. We do not propose this as "<u>the</u> way to do risk assessment"; rather it is a possible way to incorporate statistical models along with the necessary scientific judgment. We are hopeful that the day will soon come when advances in our understanding of carcinogenic mechanisms will allow the incorporation of more mechanistic approaches.

A Conceptual Model for Risk Assessment and Management

In some instances a decision tree approach can be used to categorize potential exposure situations into high, medium, or low risk. These classifications can then be used for prioritization and control strategies. There are situations, however, in which the end product of a risk assessment should be a numerical exposure limit rather than a categorization of risk. For these cases we have argued against the strict use of mathematical models because of their exclusion of relevant toxicological data.

We believe that a number of alternative approaches to risk assessment and management are possible within the framework of the four-step process advocated by AIHC (Figure 1). One possible approach combines the statistical information from the observable dose-response range with the toxicological judgments that are necessary for a logical evaluation of the data. The result is a possible formalization of the many considerations currently used by toxicologists in assessing the potential risks due to chemical exposures. We believe that this approach makes effective use of the best available data and scientific judgment.

In this section we propose a method for combining statistical and toxicological data into a quantitative framework. The result is presented in the form of a model, parts of which are quantitative and parts of which require scientific judgment. The approach taken is conceptually similar to that which has been used in other areas of toxicological risk assessment. One particular difference is that a number of specific factors are proposed; the impact of which are separately defined and evaluated. As noted earlier, we provide this concept in the spirit of stimulating discussion on this important subject and providing a direction for future work rather than the presentation of a definitive model that should be used in all cases.

This approach uses the following conceptual model to estimate the safe dose for humans.

$$\text{SAFE DOSE} = D \times \frac{1}{F} (A_1 \times A_2 \times \ldots \times A_n)(M_1 \times M_2 \times \ldots \times M_p)$$

In this model D is either a NOEL, an estimate of a NOEL, or an estimate of a response that is indistinguishable from a NOEL in any standard assay, F is an extrapolation factor based on the NOEL or E-NOEL, and A_i and M_i are factors associated with the risk assessment and risk management functions, respectively. Historically, F has been called a safety factor; however, in order to distinguish between the scientific and regulatory functions, we are calling F an extrapolation factor for it is a function of the scientifically determined best estimate of dose at some desired level of risk (starting from a NOEL or E-NOEL), as distinguished from a societally (regulatory) determined safety factor which is one of the M_i factors in the equation above.

The first step of the procedure is to use the observed dose-response data insofar as it is valid. The dose-response data can be used as a starting point for toxicological interpretation in a number of ways:

- Predict an E-NOEL (see section entitled "Alternative Analyses of Chronic Studies") by a linear or log-dose no-effect-level model, or

- Use one of three or four nonthreshold models that best fits the data (one-hit, multistage, probit, multi-hit), and extrapolate to one percent predicted tumor frequency. The models will generally agree quite well down to one percent tumor frequency; or,

- Use an experimental NOEL if it exists.

- Another approach, when the data exist, is to estimate D as a NOEL for the associated toxicity rather than tumorigenesis.

- For comparison, each of the above starting points can be modified to incorporate experimental variability in the observed data.

For most applications, the dose D estimated above will be close to the observed dose-response data and represents a limit of the dose-response information. We believe that a logical method for converting this to a safe dose for humans is to use the concept of an extrapolation factor (F) that can be modified up or down by a number of additional factors which represent information from the many types of studies that impact the total hazard potential. Selection of the particular extrapolation factors to use is a critical aspect of this approach. Other people using similar approaches have proposed values between 100 and 1000 but have tended to not separate the various components of the total extrapolation factor (US EPA, 1982; Weil, 1972; Kroes, 1982). It is important that specific components of the total extrapolation factor be independently considered both from the scientific and societal viewpoints.

The determination of A_i will be subjective and will require an agreement of expert opinion to arrive at quantification. Scientific judgment and the use of extrapolation factors is a better approach, however, than burying biological issues in mathematical sophistication. We note that if one begins with D being the dose associated with a potential risk of 10^2 (i.e., one percent), than $A_i = M_i$ for all i is equivalent to a linear extrapolation down to a dose of $10^2/F$.

Examples of "A_i factors" that may be included in specific cases are:

- Number of Positive Assays. $A_1 = 1$ if a number of well-designed bioassays are positive. $A_1 > 1$ if only one assay in a highly sensitive species is positive. For instance, the National Cancer Institute bioassay is positive in the mouse, but negative in the rat, $A_1 = 2$ may be reasonable.

- Type of Tumor $A_2 = 1$ if the compound appears to produce a diversity of tumors. $A_2 > 1$ if the only tumor is a hepatic tumor in B6C3F1 mice, or tumors of the nasal turbinate in an inhalation study of rats.

- Human Epidemiology. If human epidemiology indicates positive results, then either the risk modeling should be done from the human data, if quantitative exposure estimates exist, or $A_3 < 1$ should be used in the above approach. If human epidemiology is negative at exposures greatly in excess of anticipated future exposure conditions, then $A_3 > 1$.

- Time-to-Tumor Data. Exposure to some chemicals causes well-defined increases in tumor frequencies at an early age in animals. Other exposures result in a mixture of increases in frequency of some tumors and decreases in others and are seen only in very old animals. It may be difficult to quantify data of this type but the information is likely to be relevant for evaluating the hazard to humans. An A-factor calculated as a function of the difference or ratio between average latency and median age of death for the strain may be appropriate. For example, A_4 = Average Latency/Median Survival would result in an index that is less than 1.0 for early tumors and greater than 1.0 for old age lesions.

- Pharmacokinetic Data. A_5 can be the ratio of estimated body burdens for laboratory animals compared to humans for the same experimental administered dose (if such data exist). $A_5 > 1$ if pharmacokinetic and metabolic data indicate saturated pathways with resultant disproportionate toxicity at the doses used in bioassays with positive results.

- Approximate slope of the dose-response curve at the low end. In some cases the dose-response curve is relatively steep as it approaches the one percent level or NOEL (e.g., nitrilotriacetic acid), whereas in other cases is flatter (2-AAF, liver tumors). Although we have argued against extrapolating dose-response curves far beyond the observed data, it may be reasonable in some cases to incorporate a summary measure of slope such as the ratio of the estimated ten percent response level to the one percent response level into the safe dose calculation (i.e., ED_{10}/ED_{01}).

- Mechanism. As we understand more about carcinogenesis, the mechanism of action will become increasingly important. At the present, a logical approach would be to set $A_6 = 1$ for chemicals that are directly genotoxic and do not demonstrate any associated toxicity appropriate for predicting real or practical thresholds. On the other hand, if a specific compound has a relatively complete data base showing no in vivo or in vitro indication of interaction with DNA, and tumorigenesis occurs only at levels associated with directly observable toxicity, (e.g., cytotoxicity) than A_6 may be increasingly larger than one as the data base becomes more complete. A maximum of 10 may be appropriate.

- There are numerous other factors that, on a case-by-case basis, may be used to modify the extrapolation factor F or level of uncertainty (US EPA, 1982). These additional factors reflect special knowledge of risk potential associated with a specific situation. It is also imperative that the uncertainty associated with all estimates be communicated to the risk manager in all risk assessments.

A final step in the process is the risk management decision to modify potential human exposures upwards or downwards depending on the statutory findings required (i.e., whether the risk is significant or unreasonable) and on the impact of costs, benefits and alternatives, or other factors specified under the law. This should be done last and should be recognized as a subjective societal or political decision (i.e., risk management) rather than a scientific one. This can be done by the appropriate assignment of the risk management factors (M_1, M_2, . . . , M_p) in the calculation of the estimated safe dose for humans.

It is emphasized that although this estimate of a safe dose has been written in a simplistic equation form, we are recommending an extrapolation factor (historically called safety factor) approach in which the extrapolation factor is modified up or down depending on a number of factors. The A_i and M_i factors should be explicitly considered and judgment made on each independently. It is not possible to state specifically which factors should be included in any given risk assessment or the magnitude of the factors. These decisions must be made on a chemical-by-chemical basis. Nonetheless, we feel this is a workable framework for decision making.

This approach has many of the characteristics of that proposed by Squire (1981b); however, it is more comprehensive because it includes more aspects of the risk assessment process. It is also in keeping with recent regulatory guidelines developed in the Netherlands (Kroes, 1982) that make a clear distinction between primarily genotoxic carcinogens and primarily nongenotoxic carcinogens.

AIHC has proposed the use of one or more independent science panels to guide regulatory agencies in the scientific issues involved in risk assessment. In keeping with this concept, a science panel could recommend an appropriate extrapolation factor and could determine which A_i are different from 1, in both the positive and negative direction. The appropriate agency could then carry through the quantitative analysis as well as exposure estimation and the results would be used for appropriate risk management.

SUMMARY AND CONCLUSIONS

Our review of risk assessment and management for chemically induced carcinogenesis has focused on the use of mathematical models for the quantitative assessment of risk. We have emphasized that the risk analyst should be primarily concerned with scientific issues and the risk manager should deal with the political and social considerations. Each must understand the other's role and viewpoint. Progress is made most rapidly when each considers the other's concerns but does not let these considerations be the driving force in performing their function.

This paper focuses on the risk assessment process for we believe that good risk management decisions will result if the risk assessment is based on good science. Risk assessment is shown to be a complex, multifaceted process that is not easily quantified and is currently based on many qualitative decisions.

It is emphasized that tumor incidence rates from chronic exposure studies in animals are only one input to the risk assessment process. At this time statistical models have been developed only for these data; hence, the associated estimates of potential risk provide only one piece of the puzzle. The statistical models are useful for summarizing dose-response relationships and comparing dose-response relationships among chemicals within the observable dose-response range.

Extrapolations of tumorigenic responses to very low doses by means of mathematical models are of limited value because the models can have no validated biological basis "since these are not known for any carcinogen" (Squire, 1981b). Certainly, any extrapolations should be made with great care and only in conjunction with a variety of supporting data. To do otherwise amounts to nothing more than a blind curve fitting exercise with little predictive value.

There is also a need to develop risk assessment models that allow for differences due to species, tumor type, spontaneous incidence, sex, strain, genotoxicity, pharmacokinetics, etc. Since these important considerations as well as level of human exposure, epidemiology, and data on other toxicological end points are not presently considered by the statistical models, decisions concerning acceptable exposures should be based on a variety of inputs that include the results of statistical modeling. It is unlikely that the regulatory decision process will be even largely quantified by a single model, or group of models, in the near future.

We are not saying that the statistical approach is of no value. On the contrary, science moves forward most rapidly when the basic mechanisms can be understood, quantified, and modeled. We feel that much progress will be made when the statistical models are expanded to encompass and quantify more of the risk assessment process. Much research work must be done to accomplish this goal.

ACKNOWLEDGMENTS

The authors are pleased to thank P. F. Deisler, Jr., R. C. Barnard, and the members of the AIHC Subcommittee on Quantitative Risk Assessment, in particular, D. H. Hughes, J. T. Barr, and R. D. Bruce, for their helpful comments and suggestions on the presentation of this paper.

REFERENCES

American Industrial Health Council (1981). Chronic Health Hazards: Carcinogenesis, Mutagenesis, Teratogenesis. A Framework for Sound Science in Federal Decision Making. A statement by the American Industrial Health Council, 1075 Central Park Ave., Scarsdale, NY. Draft for external review, Oct. 30, 1981.

Anderson, M. W., Hoel, D. G. and Kaplan, N. L. (1980). A General Scheme for the Incorporation of Pharmacokinetics in Low-dose Risk Estimation for Chemical Carcinogenesis: Example — Vinyl Chloride. Toxicology and Applied Pharmacology 55:154-161.

Anderson, R. L., Alden, C. L. and Merski, J. A. (1982). The Effects of Nitrilotriacetic Acid on Cation Disposition and Urinary Tract Toxicity. Food and Cosmetics Toxicology 20:105-122.

Berenblum, I. (1929). Tumour-formation Following Freezing with Carbon Dioxide Snow. British Journal of Experimental Pathology 10:179ff.

Box, G. E. P. (1979). Robustness in the Strategy of Scientific Model Building. In Robustness in Statistics, R. L. Launer and G. N. Wilkinson, eds., pp. 201-236. Academic Press, New York, NY.

Brown, C. (1978). Statistical Aspects of Extrapolation of Dichotomous Dose Response Data. Journal of the National Center Institute 60:101-108.

Bruce, R. D. (1980). Low-dose Extrapolation and Risk Assessment. Chemical Times and Trends, October 1980, 20-23.

Carlborg, F. W. (1981a). Dose-response Functions in Carcinogenesis and the Weibull Model. Food and Cosmetics Toxicology 19:255-263.

Carlborg, F. W. (1981b). 2-Acetylaminofluorene and the Weibull Model. Food and Cosmetics Toxicology 19:367-371.

Carter, L. J. (1979). How to Access Cancer Risks. Science 204:811-816.

Cornfield, J. (1977). Carcinogenic Risk Assessment. Science 198:693-699.

Crump, K. S. (1981). An Improved Procedure for Low-dose Carcinogenic Risk Assessment from Animal Data. To be published in Journal of Environmental Pathology and Toxicology.

Crump, K. S. (1982). Designs for Discriminating Between Binary Dose-response Models with Applications to Animal Carcinogenicity Experiments. Communications in Statistics, Theory and Methods 11:375-393.

Doll, R. (1971). Age Distribution of Cancer. J. Royal Statistical Society, Series A 134:133-166.

Food Safety Council (1980). Quantitative Risk Assessment. In Proposed System for Food Safety Assessment, pp. 137-160. Food Safety Council, Washington, D.C.

Gaylor, D. W. and Kodell, R. L. (1980). Linear Interpolation Algorithm for Low-dose Risk Assessment of Toxic Substances. Journal of Environmental Pathology and Toxicology 4:305-312.

Gehring, P. J. and Blau, G. E. (1977). Mechanisms of Carcinogenesis: Dose Response Journal of Environmental Pathology and Toxicology 1:163-179.

Gehring, P. J., Watanabe, P. G. and Park, C. N. (1979). Risk of Angiosarcoma in Workers Exposed to Vinyl Chloride as Predicted from Studies in Rats. Toxicology and Applied Pharmacology 49:15-21.

Gibson, J. E. (1982). Risk Assessment Using a Combination of Testing and Research Results. In Proceedings of the Third Annual Chemical Industries Institute of Toxicology Conference, 'Formaldehyde Toxicity'; J. E. Gibson, ed. Hemisphere, New York, NY (in press).

Hoel, D. G. (1981). Carcinogenic Risk: Comment on "Regulation of Carcinogens" by E. Crouch and R. Wilson. Risk Analysis 1:63-64.

Krewski, D. (1981). Simulation of Low-dose Extrapolation Procedures. Presented at the Workshop on Biological and Statistical Implications of the ED_{01} Study and Related Data Bases, sponsored by the Society of Toxicology and National Center for Toxicological Research, Deer Creek State Park, Mt. Sterling, OH, Sept. 13-16, 1981.

Krewski, D. and Brown, C. (1981). Carcinogenic Risk Assessment: A Guide to the Literature. Biometrics 37:353-366.

Krewski, D. and Van Ryzin, J. (1981). Dose Response Models for Quantal Response Toxicity Data. In Current Topics in Probability and Statistics, M. Csorgo, D. Dawson, J. N. K. Rao and E. Saleh, eds. North-Holland, New York, NY (to appear).

Kroes, R. (1982). International Life Sciences Institute Workshop on Carcinogenic Mechanisms, Amelia Island, FL, March 16-19, 1982.

Laroye, G. J. (1974). How Efficient is Immunological Surveillance Against Cancer and Why Does It Fail. Lancet 1:1097ff.

Munro, I. and Krewski, D. (1981). Risk Assessment and Regulatory Decision Making. Food and Cosmetics Toxicology 19:549-560.

National Academy of Sciences (1980). Regulating Pesticides, Chapter 1, Report of the Committee on Prototype Explicit Analyses for Pesticides, Environmental Studies Board, Commission on Natural Resources, National Research Council.

Ramsey, J. C., Park, C. N., Ott, M. G. and Gehring, P. J. (1979). Carcinogenic Risk Assessment: Ethylene Dibromide. Toxicology and Applied Pharmacology 47:411-414.

Reitz, R. H., Gehring, P. J. and Park, C. M. (1978). Carcinogenic Risk Estimation for Chloroform: An Alternative to EPA's Procedures. Food and Cosmetics Toxicology 16:511-514.

Reitz, R. H., Quast, J. F., Schumann, A. M., Watanabe, P. G. and Gehring, P. J. (1980). Non-linear Pharmacokinetic Parameters Need to be Considered in High Dose/Low Dose Extrapolation, In Archives of Toxicology, Supplement 3: Quantitative Aspects of Risk Assessment in Chemical Carcinogenesis, pp. 79-94.

Salsburg, D. (1981). Re-examination of the ED_{01} Study — Adjusting for Time on Study. Fundamental and Applied Toxicology 1:81-87.

Schumann, A. M., Quast, J. F. and Watanabe, P. G. (1980). The Pharmacokinetic and Macromolecular Interaction of Perchloroethylene in Mice and Rats as Related to Oncogenicity. Toxicology and Applied Pharmacology 55:207ff.

Sielken, R. L., Jr. (1981). Re-examination of the ED_{01} Study — Risk Assessment Using Time. Fundamental and Applied Toxicology 1:88-123.

Squire, R. A. (1981a). Limitations of Histopathology Analysis. Presented at the Workshop on Biological and Statistical Implications of the ED_{01} Study and Related Data Bases, sponsored by the Society for Toxicology and National Center for Toxicological Research, Deer Creek State Park, Mt. Sterling, OH, Sept. 13-16, 1981.

Squire, R. A. (1981b). Ranking Animal Carcinogens: A Proposed Regulatory Approach. Science 214:877-880.

St. John, D. S., Bailey, S. P., Fellner, W. H., Minor, J. M. and Snee, R. D. (1982). Time Series Analysis of Stratospheric Ozone. Communications in Statistics, Theory and Methods, 11(12):1293-1333.

Stott, W. T., Reitz, R. H., Schumann, A. M. and Watanabe, P. G. (1981). Genetic and Nongenetic Events in Neoplasia. Food and Cosmetics Toxicology, 19:567-576.

Todhunter, J. A. (1982). Review of Data Available to the Administrator Concerning Formaldehyde and Di (2-Ethylhexyl) Phthalate (DEHP). U.S. Environmental Protection Agency Memorandum, Office of Pesticides and Toxic Substances, Washington, D.C. Feb. 10, 1982.

U.S. Environmental Protection Agency (1982). Report on Workshop on Estimating Ambient Water Quality Criteria for Epigenetic Carcinogens, Sponsored by EPA Environmental Criteria Assessments Office, 26 W. Clair St., Cincinnati, OH, Feb. 17, 1982.

Van Ryzin, J. (1981). Review of Statistics—The Need for Realistic Statistical Models for Risk Assessment. Fundamental and Applied Toxicology 1:124-126.

Wahrendorf, J. (1979). The Problem of Estimating Safe Dose Levels in Chemical Carcinogenesis. J. Cancer Research and Clinical Oncology 95:101-107.

Weil, C. S. (1972). Statistics Versus Safety Factors and Scientific Judgment in the Evaluation of Safety for Man. Toxicology and Applied Pharmacology, 21:454-463.

Weisburger, J. H. and Williams, G. M. (1980). Chemical Carcinogens, Toxicology, The Basic Science of Poisons, Casarett and Doull, eds., 2nd Ed., pp. 84-138, Macmillan, New York, NY.

Weisburger, J. H. and Williams, G. M. (1981). Carcinogen Testing: Current Problems and New Approaches. Science 214:401-407.

CHAPTER 5

GOVERNMENT DATA REQUIREMENTS FOR RISK ASSESSMENT

Dr. Joseph V. Rodricks
Environ Corporation

INTRODUCTION

The Food and Drug Administration (FDA) was the first government agency to formally incorporate risk assessment into the regulatory decision-making process. In 1972 FDA proposed to define the maximally acceptable concentration of food residues of carcinogenic drugs used in food-producing animals as that concentration which would produce a lifetime carcinogenic risk no greater than one-in-one hundred million. In its proposal the agency specified the types of data a drug's sponsor would have to develop to serve the risk assessment process, and also detailed the methodology that would be used to identify the maximal acceptable concentrations of drug residues in edible animal products. A critical feature of the proposal was the requirement that a drug's sponsor develop an analytical method capable of detecting drug residue concentrations above the maximal acceptable limit, and demonstrate that no residue of the carcinogenic drug would be found in edible products under the proposed conditions of use of the drug when such an analytical method was applied. In effect, FDA was saying that residues of carcinogens not detected with such an analytical method could be ignored without jeopardizing the health of the consuming public. Although in response to public comments FDA later changed the maximal acceptable lifetime risk to one-in-one million and modified the risk assessment methodology employed, the notion that risk assessment was a viable regulatory tool became firmly lodged.

FDA adopted this approach in response to a section of the Food, Drug and Cosmetic Act that required the agency to specify the characteristics of analytical methods used to search for carcinogenic animal drug residues in edible animal products. By adopting this approach the agency was not only taking into account the fact that carcinogens display widely different potencies (i.e., at a fixed level of risk the maximal acceptable residue levels of different carcinogenic drugs would vary according to their potencies), but was also asserting there was no need to search endlessly for carcinogenic residues in edible products taken from treated animals. Once the risk could be shown to be below a very low level, no public health benefit would be achieved by forcing it to lower levels. Although this use of risk assessment and the accompanying explicit recognition of a negligible level of risk were confined to a relatively small class of chemicals and were developed to accommodate a somewhat peculiar section of food law, they nevertheless constituted a major break with the traditional means for treating carcinogenicity data, and ushered in, for better or worse, an era in which risk assessment has become a central component of government decisionmaking.

RISK ASSESSMENT AND FEDERAL AGENCIES

The uses of risk assessment in government decisionmaking have, of course, greatly expanded since its introduction by FDA, and there are signs that its uses will increase. Risk assessment clearly fulfills an important need and, although it suffers from both a high degree of scientific uncertainty and is not always conducted and reported in a careful manner, it appears to be the only systematic means available to take into account biological information on suspect agents when deciding whether and to what extent human exposure to them should be restricted. I will now describe the essentials of risk assessment and the types of data generally regarded by federal agencies as necessary for its conduct. I will also suggest both short- and long-term improvements in risk assessment and ways in which the results of risk assessment are used in risk management decisions. For ease of presentation, and because the only important uses of risk assessment thus far concern carcinogens, I shall restrict my remarks to this class of substances.

Risk assessment is a highly uncertain enterprise. The basic problem can be stated rather simply: we have information on the risks posed by chemicals under certain conditions of exposure, but we need knowledge of the risks they may pose under other, poorly known conditions of exposure. Uncertainties arise because we are forced to make a relatively large number of assumptions and inferences to complete a risk assessment. Some of these assumptions and inferences are needed to compensate for lack of data and others are needed because of a lack of fundamental biological knowledge. Data gaps are usually chemical specific. Lack of fundamental knowledge compromises risk assessment for all chemicals.

Risk assessment is a multistep process of data analysis. When working from a moderately rich set of data, there may be more than 40 distinct analytic steps in a risk assessment. For an uncomfortably large number of these steps the risk assessor is faced with a question for which there is no clear scientific answer, but rather several plausible answers. At each such step, the risk assessor must impose assumptions or make inferences in order to complete the analysis. Typically, the assessor concerned with a carcinogen must examine at least the following questions, none of which can be answered with anything approaching scientific certainty:

- Is finding excess tumor production in experimental animals sufficient to conclude that a substance will display carcinogenic properties in humans?

- Which of several different sets of experimental animal data is most appropriate for estimating human risk?

- Which sets of tumor data should be used for high- to low-dose extrapolation?

- Which mathematical models of the dose-response curve should be used to estimate low-dose risk?

- Which measure of dose should be used for interspecies extrapolation?

- How should differences in frequency, timing, and duration of exposure be taken into account when extrapolating experimental observations to human populations?

- How should differences in route of exposure be taken into account for purposes of interspecies extrapolation?

- How should magnitude, frequency, and duration of human exposure be estimated if only limited data are available (as is almost always the case)?

- How should various exposed populations be stratified for purposes of assigning risks?

- How should final risk estimates be expressed?

For some substances there will be additional questions. Perhaps the most important of these concerns the appropriate use of those types of biological data—on pharmacokinetic and metabolic behavior of the carcinogen and on the biological mechanism by which it initiates or enhances the development of excess tumors—that reveal the true nature of dose-response relations at low dose or that assist the process of interspecies extrapolation. When data of these types are available the risk assessor must determine whether they are sufficiently complete for use in risk assessment and, if so, how they are to be used. Because all such data contain uncertainties, the risk assessor is faced with additional difficult choices whenever such data are presented.

Some of these questions (and my list is not exhaustive) can be answered in specific circumstances with a moderate or high degree of certainty if an effort is made to collect appropriate data—e.g., on absorption rates by different routes of exposure or on the nature and magnitude of human exposure. But for most of them there is little that can now be done in specific cases to narrow the range of uncertainty and thus make the risk assessor's choice less difficult.

The FDA faced these questions when the agency first introduced risk assessment and risk assessors have faced them ever since. In every case federal agencies have chosen to answer each of these questions by selecting from the various scientifically plausible options available, that option which yields the highest estimate of risk. For example, federal agencies generally assume that a finding of carcinogenicity in a single animal experiment is sufficient evidence to conclude that the substance poses a potential risk to humans. Indeed, such a finding is usually the trigger for setting into motion the risk assessment process. If there are several sets of animal and tumor incidence data, the agencies project low-dose risk from that set which yields the highest risk. They frequently use upper confidence limits on tumor incidence data. The preferred models for high- to low-dose extrapolation are those that generally yield the highest estimates of risk at low dose. The measure of dose for interspecies extrapolation is ordinarily that which either approximates or is likely to overestimate human risk. Human exposure estimation usually involves incorporation of several "worst case" assumptions and the resulting estimate of exposure is usually correct, at best, only for a small fraction of the exposed population. Other uncertainties in risk assessment tend to be similarly treated. Moreover, federal agencies generally have not incorporated any data that may shed light on underlying mechanisms of carcinogenesis of specific substances and which might thus be used to make more accurate estimates of risks than can be obtained by the application of generic models. The reason given for this last choice is that data on mechanisms are hardly ever definitive and the uncertainties in them are too great for use in risk assessment.

It should be recognized that these choices are based on what the National Academy of Sciences (NAS) has called science policy. The NAS distinguishes these types of policy choices from those that need to be made for risk management decisions. Federal agencies make science policy choices to be sure that, in the face of significant scientific uncertainty, human risk will not be underestimated. The Environmental Protection Agency (EPA) and FDA assert, for example, that their estimates of carcinogenic risks are "upper bound" estimates, and that the true risk will likely be below the upper bound and might even be zero. This can sometimes be a very wide range, but no attempt is made to define a "best estimate." The agencies also make these generic choices to ensure consistency in the risk assessment process. At least for carcinogens, the agencies do seem to have achieved a relatively high degree of consistency, although I have occasionally observed a federal risk assessment that inexplicably deviates from the norm.

If federal risk estimates are truly upper bound estimates (and I believe they are), they will tend to make individual substances appear to be more serious risks to human health than they really are. If risks estimated for a given substance are clearly in the negligible range, then the fact that they are upper bound estimates creates no problems. But if the upper bound estimates of risk for a substance appear to be of public health significance, then it is possible that the substance might be restricted to unnecessarily low levels. Is there some way to avoid this type of circumstance?

IMPROVING RISK ASSESSMENT METHODOLOGIES

There seems to me to be both short- and long-term remedies for this state of affairs. I divide them into three categories: (1) narrowing data gaps for specific substances; (2) improving the present conduct of risk assessment and the ways in which risk results are presented and used; and (3) improving the underlying methodology for risk assessment. The goal of these three remedies is to make risk assessment more accurate.

Narrowing Data Gaps for Specific Substances

The only element of risk assessment that can be readily improved in the short term is human exposure assessment. The importance of this subject and its contribution to uncertainty cannot be overstated. In my experience it is usually the single, most readily identifiable contribution to risk overestimation (although there have been cases where human risk was probably understated because of lack of appropriate exposure data).

I noted earlier that the finding of animal carcinogenicity usually prompts the regulatory risk assessment process. Once the process begins, it goes to completion, even if there are no significant data on human exposure. In such cases assumptions come into play, and these usually lead to "worst case" estimates of human exposure, both occupational and non-occupational. The affected industries are in the best position to develop and provide exposure data, and it is almost always going to be in their best interests to do so. The problem is one of priorities and timing.

It is probably safe to say that human exposure data will be needed for most compounds that are undergoing extensive toxicity testing, under the auspices of either the federal government or industry. That these tests are planned or underway is, at least for existing chemicals, hardly ever a secret. I see no reason why exposure data, including data on absorption rates by the relevant routes of exposure, should not be developed simultaneously. This seems to me a very high priority.

The only other area in which data gaps for specific compounds might be filled in a relatively short period of time is that which concerns mechanism of action. I use this term broadly to include everything from the pharmacokinetic and metabolic behavior of a substance to the reactivity of the substance or its metabolites at the cellular and subcellular level. The utility of developing this type of data is currently problematic.

In general terms, pharmacokinetic and metabolic data are probably most useful for defining the nature of the dose-response curve in the low-dose region for the animal species for which we have such data in the high-dose region. They may also be useful for assessing differences in response between humans and test species, although collecting appropriate data on this question in humans is sometimes impossible. Data on the behavior of a substance at the cellular or subcellular level may reveal whether or not a biological threshold mechanism is plausible. Other types of mechanistic investigations may similarly cast light on expected risks at low doses, and on qualitative and quantitative interspecies differences.

Although some data may be developed in a relatively short period of time and at relatively low cost, it is unlikely that answers to these questions that are fully satisfactory scientifically can be acquired without extraordinary efforts and many years of research. If regulatory agencies do not use such data in risk assessment unless they are scientifically complete—and this seems to be the present situation—then it makes little sense to develop any such data, unless there is the time and will to take the required studies to their completion (whatever that is).

This dilemma can be resolved, in part, by the second type of remedy mentioned above.

Improving the Conduct of Risk Assessment and the Ways in Which Results are Presented and Used

As I have said, federal agencies generally present upper bound numerical estimates of risk. Because results are presented in quantitative terms only, risk assessments convey the notion that the truth about risk has been uncovered. No matter how many times risk assessors state they have produced upper bound estimates, numerical estimates of predicted risk tend nevertheless to be treated with the same respect as calculated actuarial risks. This is, of course, absurd and agencies should be continually reminded of this fact.

This problem can be remedied if risk assessors resist the temptation to present only numerical estimates, but also present the sometimes substantial base of non-quantifiable information about a substance that identifies the relative merits of the various numerical estimates derived under different sets of data, assumptions, and models. Thus, careful descriptions of the strength of carcinogenicity evidence (without accompanying "yes-no" decisions about whether a substance is a human carcinogen), various test results that suggest how a carcinogen acts, relationships between carcinogen dose and pharmacokinetic behavior, and metabolic behavior of the carcinogen may provide important insights about the biological plausibility of various numerical estimates of risk. Although we rarely have complete data in these areas, this hardly seems a reason to reject what data we have. Risk assessments would benefit from a careful description and evaluation of such data, their uncertainties, and a further qualitative description of their implications for risk. Under this view of risk assessment, the risk assessor makes no a priori decisions about the amount of data and scientific certainty needed before certain facts about a substance become incorporated into a risk assessment. All data are incorporated, their implications for risk are fully described, and the relative merits and uncertainties of various numerical risk estimates are discussed.

A risk assessment of the type I envision might be difficult to use, because it requires both quantitative and qualitative results be considered. Its use requires that risk managers fully appreciate the qualitative information and how it might affect the various numerical estimates, and then provide explicit descriptions of how they incorporate all the risk information when deciding whether a risk is significant and requires reduction. The risk manager, not the risk assessor, decides how much scientific uncertainty can be tolerated. More explicit and fuller descriptions of risk would thus have the benefit of forcing risk managers to acquire better understanding of risk information and the uncertainties in it, and could also improve the risk management process because it would force risk managers to be more explicit about why they believe a risk is or is not significant.

Federal agencies might resist efforts to make changes of these types in the presentation and use of risk assessment. Indeed, there are no completely adequate models of this process that we can recommend they examine. Nevertheless, agencies should be urged to move in this direction, at least in those cases where efforts have been made to collect more telling biological information than is obtained in the typical bioassay.

Improving Risk Assessment Methodology

The third type of remedy for the problems of risk assessment is the most ambitious and is the most difficult to discuss. Currently used experimental models for collecting toxicity data do not, of course, yield the truth about human risk and it is lack of fundamental biological knowledge that is the major bar to this truth. Although the methods of epidemiology are the only means available to gain the truth about how environmental agents affect human health, there are so many difficulties in collecting and evaluating epidemiological data that we shall no doubt continue to use experimental models as our primary information gathering device. The question then becomes whether we are using—or even thinking about using—the correct models.

I suggest most of our current toxicity test methods do not include a means for attaining true knowledge about the risks of environmental agents, and we are thinking little about how current knowledge and techniques might be incorporated into the tests we now perform to gain information of clear and direct relevance to the major uncertainties in risk assessment procedures. Nor are we thinking much about the types of research efforts required to fill these gaps.

The issue of appropriate models for high- to low-extrapolation, for example, is one of several major areas of uncertainty in risk assessment. No animal test or other experimental models we now use reveals any information about the form of dose-response relations in the low-dose range typically of interest. Clearly, we are not going to gain such information unless we measure endpoints other than tumors or other gross manifestations of toxicity. But are there not other endpoints—biochemical or physiological—that may be correlated with tumor development or other forms of toxicity whose measurement might greatly improve sensitivity of bioassays and reveal highly useful extensions of dose-response curves? I am not sure this approach is the best one, but I see little concerted thinking and almost no research about how to approach this enormously important problem.

SUMMARY AND CONCLUSIONS

The time seems right for a substantial effort to incorporate as much new technology for studying the behavior of chemicals in animals and in other experimental systems into the various toxicity test systems we now use. However, this effort should be based on knowledge of the uncertainties in risk assessment methodologies so that we can determine which of these newer experimental technologies best fill the gaps in risk assessment and how the results of their use will be incorporated. It makes little sense to incorporate new methodologies without knowing how we shall use the results they produce. I suspect such an effort will reveal the need for fewer chronic, whole animal studies and will lead to the development of more telling toxicity data in less time and at lower costs. The sort of effort I envision could also, of course, help identify the major gaps in risk assessment for which no technology now exists, and ideally would suggest the various types of research efforts that might most usefully fill these gaps.

I suggest these objectives can be achieved by gathering together those scientists most in touch with the best available methods for assessing the biological behavior of chemicals in experimental systems, those government scientists and officials who are required to specify the types of toxicity tests needed to fulfill regulatory requirements, and those scientists who use information from such tests in assessing risks. Such a gathering, held under both government and industry sponsorship, would not simply be a symposium, but a true working conference with expert papers developed in advance containing specific proposals for achieving improvements in each of the important elements of risk assessment and for guiding the nation's research agenda in these areas. Such a working conference, if conducted properly, would have a high probability of inspiring the long-sought consensus on toxicity test requirements and risk assessment methodology and might even affect the thinking of research scientists.

Until we make moves in these directions, we shall continue to develop and use extremely costly experimental toxicity data for the relatively narrow purpose of standard-setting. It seems time to develop and use experimental models that not only serve the standard-setting process, but that also ensure that the standard-setting process is based on more accurate estimates of human risk. We would thereby gain vital knowledge about the true seriousness of the threat to human health posed by environmental agents which is, after all, the ultimate objective.

CHAPTER 6

UTILIZATION OF RISK ASSESSMENT IN CORPORATE RISK MANAGEMENT DECISIONS

Dr. Paul F. Deisler, Jr.
Shell Oil Company

INTRODUCTION

Many companies have ways of evaluating and abating health risks when existing regulations are inadequate or when no regulation exists. Evidence of these activities are the citations found, for example, in Occupational Safety and Health Administration's (OSHA) Notice of Proposed Rulemaking for ethylene oxide.1/ Some companies have made their systems public; it is hoped more will do so in time.

1/ Occupational Safety and Health Administration (OSHA), "Occupational Exposure to Ethylene Oxide," Federal Register, XLVIII, (1983) pp. 17, 284-317, 319.

EVALUATING RISK

I would like to discuss Shell Oil Company's systematic way for evaluating and abating risks.[2/] I will focus on cancer, an important interest, but the system is applicable to other hazards. It has been developed gradually, over the last several years, and applied in an early form as far back as five years ago. Its first rigorous application was to ethylene oxide.

The system consists of:

- a process composed of several orderly, sequential steps in which basic information is analyzed, risks assessed, and actions (if any) are decided upon;

- a quantitative yardstick for use where quantitative risk assessment is possible to help distinguish between higher risks, lower risks, and insignificant risks. The extreme uncertainty in the data and the bluntness of our interpretive tools allow us no more than three levels;

- a philosophy outlining appropriate action on the basis of the risk level found, not only where quantitative risk assessment is possible, but where qualitative risk assessment is possible; and

- management mechanisms for using these risk assessment tools.

2/ P. F. Deisler, Jr., "Dealing with Industrial Health Risks: A Step-Wise, Goal-Oriented Concept," American Association for the Advancement of Science Special Symposium No. 65: Risk in a Technological Society, ed. by C. Hohenemser and J. X. Kasperson (Boulder, CO: Westview Press, 1982). Also, Deisler, "Fundamental Problems and Practical Solutions in Assessing and Abating Risks that Chronic Chemical Exposures Pose for People," (Paper presented at the meeting of the American Association for the Advancement of Science, Santa Barbara, CA, June 23, 1982). Also, P. F. Deisler, Jr., J. E. Berger, and R. L. Brunner, "A Systematic Approach to Reducing the Risk of Industrially Related Cancer," Reg. Toxicol. and Pharmacol., III (1983), pp. 26-27.

Process

The intent of the process is to ensure scientific analysis is kept separate from the later analysis of what to do about a risk when a risk requiring abatement exists. A number of other organizations, governmental and private, have developed and published their own risk assessment and management processes aimed primarily at governmental regulatory systems.[3/]

In my company's process, we identify four principal steps as seen in Figure 1. Beginning with the question of whether or not a hazard exists, data are gathered and validated in the step known as hazard identification. The intent of this step is to determine which data are applicable and valid and whether they indicate the existence of a hazard, where hazard is the potential to cause harm. If a hazard is indicated, the process proceeds to hazard evaluation, a twofold step: a qualitative evaluation first, and then, if possible, a quantitative evaluation. Quantitative evaluation is based on identification of applicable data provided during qualitative evaluation. Ideally, the intent of this step is to develop ranges of responses, qualitative and/or quantitative, of humans to laboratory-like exposure conditions, based on animal data, epidemiologic data, or a combination of these and other ancillary information. The ideal can be, at best, only roughly approximated.

3/ D. R. Calkins, R. L. Dixon, C. R. Gerber, Dr. Zarin, and G. S. Omenn, "Identification, Characterization, and Control of Potential Human Carcinogens: A Framework for Federal Decision-Making," Journal of the National Cancer Institute, LXIV, No. 1 (1980), pp. 169-176. Also, P. F. Deisler, Jr., "Science, Regulations and the Safe Handling of Chemicals," Reg. Toxicol. and Pharmacol., II (1982), pp. 335-344. Also, European Chemical Industry Ecology and Toxicology Centre (ECETOC), Risk Assessment of Occupational Chemical Carcinogens, (Monograph No. 3, ECETOC, Brussels, Belgium, January, 1982). Also, Food Safety Council, Proposed System for Food Safety Assessment, (Final Report of the Scientific Committee of the Food Safety Council, Washington, D.C., June, 1980). Also, Federal Cosmetic Toxicology, XVIII (1980), pp. 711-734. Also, National Research Council, Risk Assessment in the Federal Government: Managing the Process, (Washington, D.C.: National Academy Press, 1983).

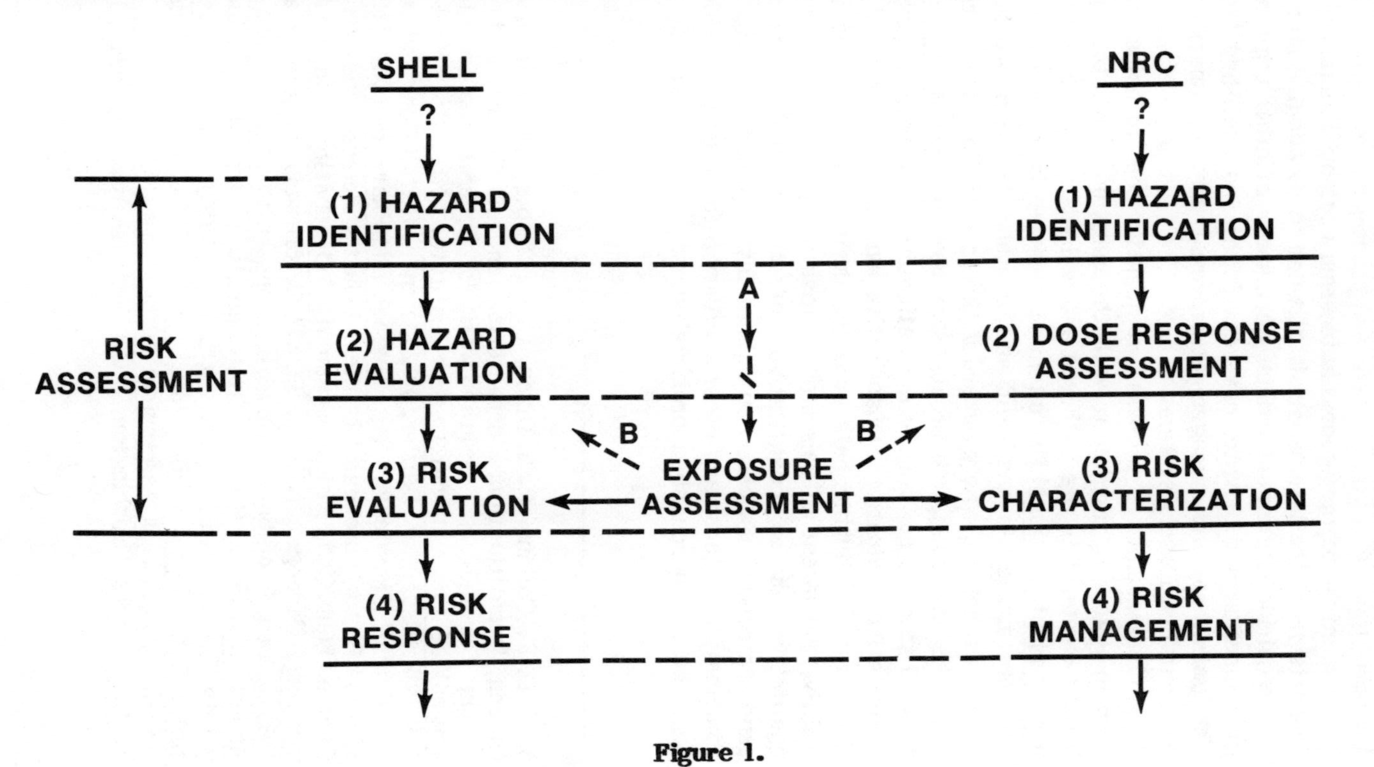
RISK ASSESSMENT AND RESPONSE PROCESSES
SHELL
?
(1) HAZARD IDENTIFICATION
(2) HAZARD EVALUATION
(3) RISK EVALUATION
(4) RISK RESPONSE
NRC
?
(1) HAZARD IDENTIFICATION
(2) DOSE RESPONSE ASSESSMENT
(3) RISK CHARACTERIZATION
(4) RISK MANAGEMENT
A
B
B
EXPOSURE ASSESSMENT
RISK ASSESSMENT

Figure 1.

While these first two steps are proceeding, a parallel exposure assessment is begun (arrow A). In this activity the routes, levels, and patterns of exposure are determined and evaluated so they may be used to lesser degree in the second step (arrow B), but intensively in the third or risk evaluation step. In this latter step the results of the second step, hazard evaluation, are used with exposure assessment to determine the level of risk. The first three steps are what are commonly called risk assessment and the end product is a characterization of the risk. Based upon that characterization and assuming that the risk warrants action, the fourth step, risk response, is employed in which measures to reduce or abate the risk are decided. Once this is complete, risk abatement measures are then implemented.

Each of the steps described is a decision step. Together, decisions made at each step and the characterization of risk determine resources allocation and prioritization of effort so the most important risks can be handled first.

Alternatives considered in the risk response step include: setting internal standards, modifying operating methods, changing plant design, imposing administrative controls, utilizing personal protection, modifying maintenance procedures or distribution or packaging methods, considering alternative processes or process changes, eliminating certain end uses, or eliminating a product from manufacture and sale. In addition to this, it is necessary to inform and possibly to train workers. It is also our practice to inform customers when we find a risk which causes us to take measures over and above the requirements of existing regulations.

Quantitative Yardstick

The quantitative yardstick used for characterizing risk defines the boundaries between the three regions of risk: higher risk (HR), lower risk (LR), and insignificant risk (IR). The boundaries are defined quantitatively in terms of probability, p, where the probability is that of an animal or human suffering at least one adverse effect—in the case of cancer, at least one tumor. This probability or risk must be reduced, whether a specific exposure situation contains one, a hundred, or more carcinogens. This reduction is brought about by controlling exposures.

The boundaries between the risk regions are designated as follows: for the boundary between HR and LR, the Level of Action

(LOA); and between LR and IR, the Level of Insignificance (LOI). In this system we thus deal additionally with an upper level of risk as opposed to dealing only with a de minimis risk as is often the case in most risk considerations in the regulatory sphere. In this way, setting priorities is facilitated.

In setting levels of p for the LOA and the LOI, levels of risk already considered in other serious cases are of interest. In the workplace, for example, the risk of accidental death of a worker who is on the job for 30 years[4/] is approximately one chance in four hundred (2.5×10^{-3}). The Supreme Court in the Benzene case[5/] considered a risk of one in one thousand (10^{-3}) as being high enough to warrant reduction or elimination. In a number of other cases, lower levels of lifetime risk have been considered as being insignificant or de minimis, or at least low enough to warrant no action.[6/] Some of these risks are 10^{-5} (risk targets for criteria pollutants under the Clean Water Act), 10^{-6} (for indirect additives to food under the Food, Drug and Cosmetic Act), or even zero for direct food additives (under the Delaney Clause of the Food, Drug and Cosmetic Act). A recent paper[7/] considered what levels of risk

---

4/ R. Wilson, Direct Testimony Before the United States Department of Labor Assistant Secretary of Labor for Occupational Safety and Health Administration, OSHA Docket No. H090, Washington, D.C., 1978.

5/ United States Supreme Court, Industrial Union Department, AFL-CIO v. American Petroleum Institute, et al., No. 78-911, decided July 2, 1980.

6/ T. C. Byerly, USDA Policy on Carcinogens, (Contract 43-32R7-9-1100, July 2, 1979). Also, Environmental Protection Agency (EPA), "Requests for Comments on Water Quality Criteria for 27 Toxic Water Pollutants," Federal Register, XLIV, No. 15 (1979), p. 926. Also, Food and Drug Administration (FDA), "Chemical Compounds in Food-Producing Animals," Federal Register, XLIV, No. 17, (1979), p. 70.

7/ J. G. Cobler and F. D. Hoerger, "Analysis of Agency Estimates of Risk for Carcinogenic Agents," Proceedings of the Symposium on Risk Analysis in the Private Sector, ed. by V. T. Cavello and C. Whipple, Third Annual Meeting of the Society for Risk Analysis; New York: Plenum Press (in press) 1983.

have, in fact, been attained in agency actions in the past. OSHA's risk distribution was found to have a mode close to 10^{-4}, whereas other agencies' modes tended to be somewhat below that. In both cases the distributions of risk were highly skewed, with the modes being higher than the means.

A factor to consider in setting the level of risk for LOAs for different types of exposure situations is the degree to which the risks are voluntarily or involuntarily taken. Between risks that are voluntarily taken and those which are involuntarily taken there can be as great a difference as a factor of approximately one thousand (three orders of magnitude) in acceptance.[8/] Considering risk in the workplace against risks posed by air pollution or water pollution, for example, risks in the workplace may be considered semi-voluntary compared to air or water pollution, which are essentially involuntary. In the workplace, the trained and informed worker has some ability to control the degree of risk. Differentiation on this basis is thus possible between the workplace and other types of exposure situations amounting perhaps to one to two orders of magnitude in risk acceptance.

As far as types of exposure situations are concerned, there are a small number of these. I mentioned three. Others include indirect food additives, consumer products, and medicinal drugs. Other types of exposure situations may perhaps be defined in the future. Every actual exposure situation, however, is a specific one; that is to say, the exposure to a specific workplace, to water from a particular drinking water supply, and so forth. Distinguishing between types of exposure situations (few in number) and specific exposure situations (many in number) is important in later utilization of the probability levels corresponding to the LOA and the LOI.

In addition to considerations already enumerated, setting a goal for risk reduction is useful in selecting the probability corresponding to the LOA for any given type of exposure situation. Goal-setting is an accustomed practice in many other kinds of activities. For example, setting safety goals to decrease incidence levels helps deal with uncertainty.

8/ C. Starr, Perspectives on Benefit-Risk Decision Making, National Academy of Engineering, 1972.

Recent cancer research indicates the workplace alone contributes in the range of one to five percent, or perhaps two to eight percent of total cancer[9/] with other sources of industrial pollution contributing additional percentage points. Setting a goal to reduce industrial source contribution to one percent would, if achieved, be a signal improvement over today's situation. Moreover, the possibility of measuring that amount of contribution would be very small indeed. That is, the total contribution of industrial pollutants would be so low, it would be within the "noise level" of our ability to determine the contribution.

The LOA can be determined for each type of exposure situation using indicated parameters without having to refer to such specifics as numbers of people affected. The LOI, on the other hand, will be sensitive to numbers of people affected and to background cancer rates for each specific exposure situation. I will focus first on selecting the LOA which, because it is the dividing line between HR and LR, is the more important of the two boundaries.

Figure 2 summarizes relationships important in setting and utilizing the LOA. The first expression under "Criteria" is the most basic expression. On the right-hand side the term p_{ci} represents a ceiling value of probability, a probability corresponding to the LOA for the i-th type of exposure situation. The term f_i is the fraction of the total population exposed to the particular type of exposure situation. The summation indicates the summing of the products of these fractions and probabilities over all k types of exposure situations. If individuals exposed to each type of exposure situation were to be subjected to probabilities at these ceiling levels, the average joint probability of cancer among the full population (from industrially related agents) would be given by the right-hand side of the expression. The ceiling probabilities are, however, simply probabilities; they are not related to or connected to any actual physical agents, but represent standards of risk. Therefore, they may be summed as shown, provided they are small enough so no significant

---

9/ R. Doll and R. Peto, The Causes of Cancer, (Oxford: Oxford University Press, 1981). Also, J. Higginson and C. S. Muir, "Environmental Carcinogenesis: Misconceptions and Limitations to Cancer Control," Journal of National Cancer Institute, Vol. LXIII (1979), pp. 1291-98.

CRITERIA:

$$(1) \quad g_c F = \sum_{i=1}^{k} f_i p_{ci}$$

$$(2) \quad p_c = \sum_{i=1}^{k} p_{ci}$$

WHERE:

$$\sum_{i=1}^{k} f_i \leqslant k$$

OBJECTIVES:

$$(1) \quad p_i < p_{ci}, \quad [1 \leqslant i \leqslant k]$$

$$(2) \quad g < g_c$$

Figure 2.

error occurs. Their sum is actually an estimate of the average, joint probability of the k statistically independent ceiling probability values.

The left hand side of equation (1) under "Criteria" in Figure 2 contains two terms: F, which is the fraction of the total population expected to contract cancer in their lifetimes from all causes (0.25), and g_c which is the ceiling value of the fractional contribution of industrial sources of pollution to total cancer that we wish to impose as a goal to be achieved or improved (for example, one percent or 0.01). Since the left-hand side of the equation can be established by choosing g_c and by knowing F and the f_i on the right-hand side are either known or can be estimated easily, it is a matter of making choices to select a set of p_{ci}-values to conform to the various criteria already discussed. Setting all the f_i equal to 1.0, and for k = 6, and allowing a factor of 20 (1.3 orders of magnitude) to exist between the workplace and the other five types of exposure situations named above (that is, the ceiling probability values for the other five are one-twentieth of that for the workplace), it is calculated that the ceiling probability for the workplace is 2×10^{-3} and that for the other five types of exposure situations the P_{ci}-values are all 10^{-4}. This is somewhat conservative since it is not true that 100% of the population is exposed equally to all types of exposure situations. Especially in the case of the workplace, the fraction is less than 100%. For example, in the case of the workplace, assuming 25% of the population were exposed while all other factors except g_c remained the same in the equation, the corresponding value of g_c would be 0.004 or 0.4%, compared with the one percent value selected in the first place.

The second expression under "Criteria" merely states that we should assure ourselves that those individuals who are exposed 100% to all types of exposure situations do not exceed a risk greater than a grand ceiling, p_c. The inequality shown just below expression (2) merely indicates that for any individual the sum of the f_i values can be as large as k, the total number of the types of exposure situations (in this case, k is 6).

Expression (2) at the bottom of the figure under "Objectives" states the goal: The actual contribution of industrially related cancer is to become <u>less</u> than g_c. For actual probabilities which are statistically independent, expression (1) under "Objectives," when met in actuality, will accomplish the major objective of expression (2). If synergism between agents <u>within</u> a specific exposure situation is encountered, but statistical independence obtains between specific exposure situations, Objective (1) can be met although the difficulty here is to determine what the synergistic law might be. However, since the goal is to fall below ceiling values as stated in Objective (1), reducing exposures as much as reasonably possible achieves some margin of safety against the eventuality of synergism, error, or other unknowns and uncertainties.

A more difficult case is that of synergism between agents which occur in different exposure situations to which the same individuals are exposed. Here, too, the practical approach is to assume independence and to control exposures as low as reasonably possible to provide a margin of safety and not just as low as necessary to meet exact ceiling values of probability. In either case, it is desirable to establish programs of monitoring, medical surveillance, and ongoing epidemiologic studies to discover whether the goals have in fact been attained. While this means goals may have to be attained stepwise, it also means improved safety is imposed at an early date while determining what further requirements may be needed. At present, the state of our knowledge and understanding permits us to do no better.

Considering the numerical values suggested for the LOAs and the criteria suggested for establishing these, the suggested LOAs are reasonable. This is not to say, however, that other sets of numbers falling within the kinds of ranges of probabilities already discussed might not also be sound. There is an element of choice in setting the LOAs.

Figure 3 shows the LOA and LOI as they relate to each other for a single type of exposure situation. Across the top of the figure an extensive variable, E, is plotted; for purposes of this discussion consider E to be log N, where N is the number of individuals in a specific exposure situation. Thus, in this figure, the numbers of people involved in specific exposure situations of this type of exposure situation range from one to ten billion. As stated earlier, the LOA is constant, having been selected by a combination of goal setting, comparative risk determinability and feasibility (compare with the results of Cobler and Hoerger.[10/] The LOI are shown as

10/ J. G. Cobler and F. D. Hoerger, "Analysis of Agency Estimates of Risk for Carcinogenic Agents," Proceedings of the Symposium on Risk Analysis in the Private Sector.

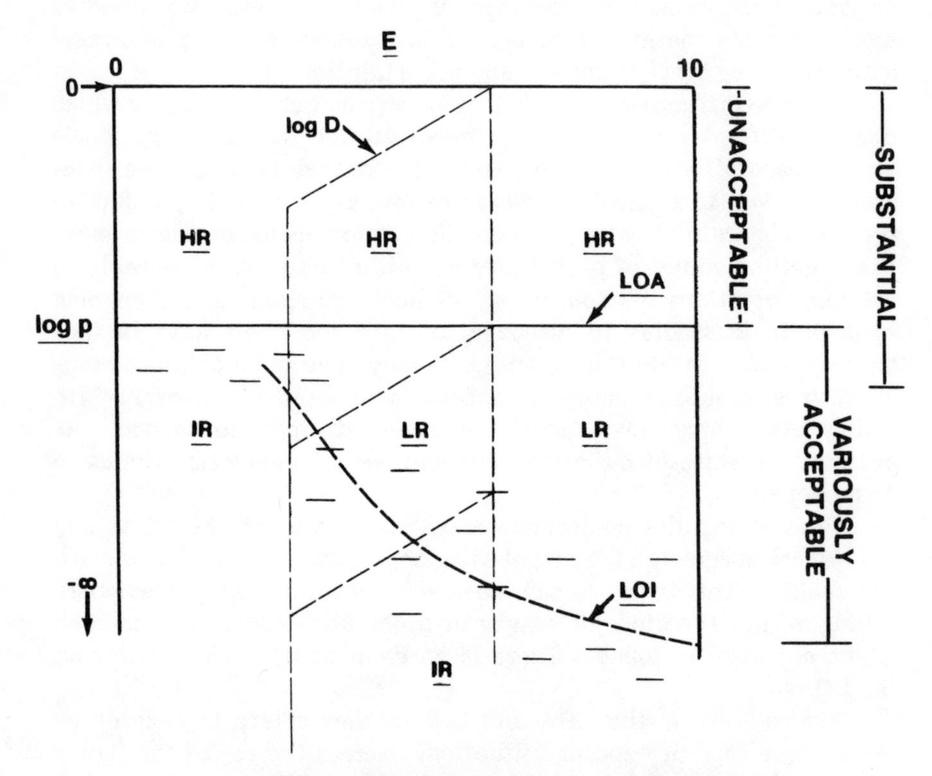

THREE RISK REGIONS AS RELATED TO E FOR A SINGLE TYPE OF EXPOSURE SITUATION (IN TERMS OF p, ONLY)

Figure 3.

individual dashes surrounding the dashed trend line. Even if the LOI were established on purely statistical grounds, the fact that different numbers of people are involved and, for a single value of N, the fact that the background levels of cancer will differ from one specific exposure situation to another, will lead to the kind of situation depicted in this figure. At the left-hand side of the figure, the LOI approaches the LOA; indeed, LOI-values higher than the LOA could in principle be calculated but, because the LOA represents a ceiling, such LOI-values are not accepted. For situations with small numbers of people it may be convenient simply to establish a margin (for example, a factor of ten), between the LOA and the LOI.

At the right-hand side of the figure, the regions are compared with a number of terms. The term "unacceptable" in this context is not a societal term, but one which simply implies the goal will not be attained if risks in the HR region are allowed to exist. On the other hand, risks between the LOA and the LOI are not necessarily fully acceptable (that would correspond to risks below the LOI), but are instead "variously acceptable." For those used to considering the possibility of making reports under section 8(e) of the Toxic Substances Control Act, I like to think the term "substantial" overlaps HR completely and reaches part way into LR; this is a simple, qualitative way of thinking about risk.

The LOA, the LOI, and the three risk regions, HR, LR, and IR, are shown in Figure 4. This plane corresponds to the plane outlined in dashed lines in the previous figure. Superimposed are a series of lines which together form a display of the ranges of quantitative responses, which our analysis has indicated might represent the human response.

This display is typical of those developed during our ethylene oxide risk assessment. The solid line marked A represents the highest extrapolation of actual data to lower exposure levels including the highest translation of dose to allow for scaling from the animal to the human. The lower line marked B represents the lowest such extrapolation and translation. As is well known, several of the various extrapolation functions that might represent the data[11/]

11/ Food Safety Council, Proposed System for Food Safety Assessment. Also, C. N. Park and R. D. Snee, "Quantitative Risk Assessment: State-of-the-Art for Carcinogenesis," Fundamental and Applied Toxicology, III (1983), pp. 320-333.

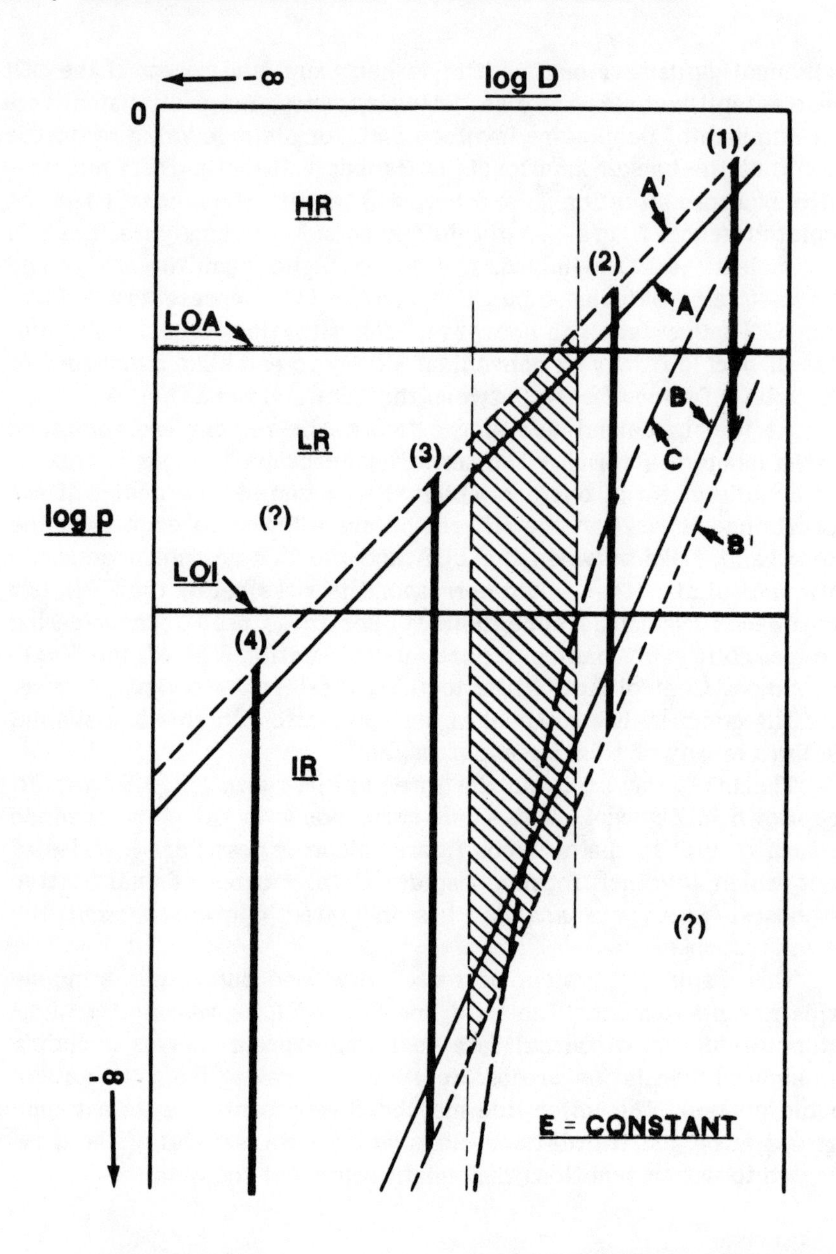

PROBABILITY (p) — DOSE EXTRAPOLATION

Figure 4.

frequently give a good fit within the experimental range and so offer no reasons for excluding any. This plot, then, shows the region within which all these extrapolations lie, with exposure corrected for scale using the several usual body-size factor methods. The lines A' and B' indicate the upper and lower confidence levels of A and B, respectively.

Clearly, the expected responses are not pinpointed, but cover a wide range. For example, the vertical solid lines numbered (1) through (4) indicate the ranges of responses expected at exposure levels corresponding to those dose levels. Line (1) lies largely in the HR region and describes the risk as lying in HR; line (4) lies in the IR region and the risk, therefore, is described as lying in that region. Lines (2) and (3) offer somewhat greater problems, but if they are judged to lie within the LR region and the risks are then abated accordingly, no serious error will be made. In actual fact, the distribution of exposures is such that one will normally not have simple vertical lines such as those shown, but something more like the shaded area. This poses no special problem since the logic for deciding which risk region is the appropriate one is much the same.

Since the extrapolation functions commonly used have no specific relationship to individual causes of carcinogenesis, one must always bear in mind that establishing risks by this means is an uncertain business; indeed the true function might well be something like the dashed line, C, for example. For this reason, question marks are shown in Figure 4 to indicate that, while one may think the risks lie within the boundaries indicated, there is some probability that such is not the case. Obviously, the final characterization of risk in examining displays of this sort and in considering all assumptions made in arriving at such displays, is judgmental and must always be accompanied by a healthy degree of skepticism and common sense when weighed with other, qualitative evidence or when deciding what action to take, if any. In the management of the risk assessment process, it is useful to establish a standing committee or team of experts who, with an efficient and skilled chairman, are accustomed to working with the system and with each other.

Philosophy

In the final analysis, the philosophy for dealing with risk once it has been decided in which region risk lies is as follows:

- If the risk lies in the HR region, to meet the goal it is necessary to take actions of the types indicated to move out of that region.

- If the risk lies in the LR region, the risks must be reduced in a reasonable way to a lower level. This applies to risks which have already been reduced from HR to LR as well as to risks found to lie in the LR region in the first instance.

- Finally, risks found to lie in the IR region need not be further reduced. In this case, monitoring systems must be established and maintained to assure that the risk remains insignificant.

In the event that quantitative analysis is not possible, even in the sense used in this method, a qualitative analysis may nonetheless be possible and may lead to the conclusion that a risk exists requiring reduction though it may not be possible to distinguish between risks lying in regions HR and LR. In this case, the recommended procedure is to treat the risk as though in region LR and reduce it as much as reasonably possible while gathering the data and the further information to determine where the risk lies. In this way, reasonable steps are taken to reduce risk so that, if it is found the risk was indeed in the HR region to begin with, it will have been at least partially mitigated by the time this discovery is made and further mitigation, if necessary, can then be applied.

It is an explicit feature of this system to recognize the fact that we cannot have all knowledge immediately nor can we know exactly how to proceed in each case to reduce risk. We must move in a stepwise fashion, modifying our views as new information is obtained and interpreted. Figure 5 shows the overall objectives this system is intended to accomplish. The first statement says the intention is to deal with highest priority risks in the HR region and to take steps deploying the resources and information available to reduce those risks. The second statement makes it clear that, having taken steps to reduce risks, we may not have achieved the most that can be achieved. Therefore, steps must be taken to permit the longer term evaluation of whether the risks indeed have been reduced and, if not, to permit us to continue to reduce future risks. Finally, in keeping with the goal and some of the criteria with which we have established the LOA values, it is our desire to reduce

CANCER RISK REDUCTION
A GOAL

(1) DEPLOY RESOURCES AND INFORMATION SO AS TO PROTECT THE MOST PEOPLE FROM THE RISK OF CANCER AS SOON AS AND AS EFFECTIVELY AS POSSIBLE.

(2) LAY A FOUNDATION FOR CONTINUING RISK REDUCTION AND CONTROL.

(3) REDUCE THE CONTRIBUTION OF CANCER FROM INDUSTRIALLY DERIVED AGENTS TO TOTAL CANCER TO AN INSIGNIFICANT LEVEL.

Figure 5.

the contribution of cancer from industrially derived agents to a truly insignificant level. The LOA-values themselves should be subject to occasional review and the inadvertent raising of these risk ceilings (for example, if F should rise) must be avoided.

SUMMARY AND CONCLUSIONS

As our understanding of the science of cancer improves, our ability to utilize this kind of system will also improve. However, a new challenge now asks our industry to deal with risks which, according to our knowledge, lie entirely within the IR or insignificant risk region. For example, the present concerns and expressions of fear at the occurrence of dioxins at levels of a few tens of parts per quadrillion in water or other similar cases are causes for real concern. Here we are asked to deal with risks that probably pose no real health risk, but which are nonetheless considered worthy of attention and of the serious expenditure of efforts of talented people because of the generation of public fear. This is a chilling thought, and when we speak of trying to "reduce" risks the total elimination of which would have inappreciable effects on public health, we are facing an almost impossible problem. Science, the political process, and legislative reform can help if they work together successfully. This is the most serious challenge we face in the whole field of risk of adverse health effects.

BIBLIOGRAPHY

Byerly, T. C., USDA Policy on Carcinogens, U.S. Department of Agriculture, Office of the Secretary, USDA ad hoc Departmental Policy Group on Carcinogens, USDA Contract 43-32R7-9-1100, July 2, 1979.

Calkins, D. R.; Dixon, R. L.; Gerber, C. R.; Zarin, D.; and Omenn, G. S. "Identification, Characterization, and Control of Potential Human Carcinogens: A Framework for Federal Decision-Making." Journal of the National Cancer Institute, Vol. LXIV, No.1 (1980), 169-176.

Cobler, J. G. and Hoerger, F. D. "Analysis of Agency Estimates of Risk for Carcinogenic Agents." Proceedings of the Symposium on

Risk Analysis in the Private Sector, Third Annual Meeting of the Society for Risk Analysis, New York. Edited by V. T. Covello and C. Whipple. New York: Plenum Press (in press), 1983.

Deisler, P. F., Jr. "Dealing with Industrial Health Risks: A Step-Wise, Goal-Oriented Concept." American Association for the Advancement of Science Special Symposium No. 65: Risk in a Technological Society. Edited by C. Hohenenser and J. X. Kasperson. Boulder, CO.: Westview Press, 1982.

Deisler, P. F., Jr. "Fundamental Problems and Practical Solutions in Assessing and Abating Risks that Chronic Chemical Exposures Pose for People." Paper presented at the Meeting of the American Association for the Advancement of Science, Santa Barbara, CA, June 23, 1982.

Deisler, P. F., Jr. "A Goal-Oriented Approach to Reducing Industrially Related Carcinogenic Risks." Drug Metabolism Reviews, Vol. XIII, No. 5 (1982), 875-911.

Deisler, P. F., Jr. "Science, Regulations and the Safe Handling of Chemicals." Reg. Toxicol. and Pharmacol., Vol. II (1982), 335-344.

Deisler, P. F., Jr.; Berger, J. E.; and Brunnel, R. L., "A Systematic Approach to Reducing the Risk of Industrially Related Cancer." Reg. Toxicol. and Pharmacol., Vol. III (1983), 26-27.

Doll, R., and Peto, R. The Causes of Cancer, Oxford: Oxford University Press, 1981.

Environmental Protection Agency. "Requests for Comments on Water Quality Criteria for 27 Toxic Water Pollutants." Federal Register, XLIV, No. 15 (1979), 926.

European Chemical Industry Ecology and Toxicology Center (ECETOC). Risk Assessment of Occupational Chemical Carcinogens. Monograph No. 3, Brussels, Belgium, January 1982.

Food and Drug Administration. "Chemical Compounds in Food-Producing Animals." Federal Register, XLIV, No. 17 (1979), 070.

Food Safety Council. Proposed System for Food Safety Assessment, Final Report of the Scientific Committee of the Food Safety Council, Washington, D.C., June, 1980. (See also Federal Cosmetic Toxicology, Vol. XVIII (1980), 711-734.)

Higginson, J. and Muir, C. S. "Environmental Carcinogenesis: Misconceptions and Limitations to Cancer Control." Journal of the National Cancer Institute, LXIII (1979), 1291-98.

National Research Council. Risk Assessment in the Federal Government: Managing the Process, Washington, D.C.: National Academy Press, 1983.

Occupational Safety and Health Administration (OSHA). "Occupational Exposure to Ethylene Oxide." Federal Register, Vol. XLVIII, pp. 17, 284-317, 319 (1983).

Park, C. N. and Snee, R. D. "Quantitative Risk Assessment: State-of-the-Art for Carcinogenesis." Fundamental and Applied Toxicology, Vol. III (1983) 320-333.

Starr, C. Perspectives on Benefit-Risk Decision Making. Committee on Public Engineering Policy, National Academy of Engineering, Washington, D.C., 1972. (Available through National Technical Information Service, P.B. 213685.)

United States Supreme Court. Industrial Union Department, AFL-CIO v. American Petroleum Institute, et al. No. 78-911, decided July 2, 1980.

Wilson, R. Direct Testimony Before the United States Department of Labor Assistant Secretary of Labor for Occupational Safety and Health Administration, OSHA Docket. No. H090, Washington, D.C., 1978.

CHAPTER 7

REMARKS TO THE CHEMICAL MANUFACTURERS ASSOCIATION

Senator David Durenberger
Committee on Environment and Public Works

INTRODUCTION

I have three items to discuss with you. The first is my general philosophy on risk assessment and management, the theme of your conference. The second is a brief preview of the agenda for my environmental subcommittee in 1984 including reauthorization of the Toxic Substances Control Act (TSCA). Finally, I have a few suggestions for your industry—steps you might take to deal successfully with the difficult public policy issues you face.

PUBLIC POLICYMAKING IN RISK MANAGEMENT

Risk assessment. Risk management. It has become fashionable to separate the consideration of risk into two parts. Risk assessment—the estimation of the association between exposure to a substance and the incidence of some disease—is presumably a job for the scientist. Risk management is the process of deciding what to do about the risk once determined. It calls upon many disciplines. When the risk affects human health or the environment, the job usually lands on the desk of a public policymaker.

This separation has become fashionable, I suspect, because none of us are happy with our choices about risk. The policymaker is not happy with the carefully qualified answers he gets from the scientist. The scientist is not comfortable weighing the costs and benefits that must be included in every choice of the policymaker.

In a recent issue of Science magazine, I read about a third discipline—something called "science policy." According to the authors of this article concerning the formaldehyde decisions made by the Environmental Protection Agency (EPA), "science policy" denotes issues grounded in scientific analysis, but for which technical data does not clearly support one conclusion or the other. I think the authors wanted to turn the job of "science policy" over to the courts.

We are all in the business of science policy. However good they make us feel, these distinctions cannot hide the reality that policy and science go hand-in-hand through every step in making choices about risk. We must not lose sight of the fact that risk assessment is far from a purely scientific endeavor. For example, there are several cancer models for extrapolating from high-dose animal tests to low-dose human effects and, for the same set of test results, scientists give estimates of cancer incidence that will vary by several orders of magnitude. There is no irrefutable scientific basis for selecting among these models, so the scientists' choice is easily influenced by his or her policy preferences for risk aversion.

And it is sciences that makes management possible. Even when the policymaker knows precisely what risk he or she will accept for a specific kind of activity, getting there is not simply a matter of wishing it so. Quite often it means stretching science and engineering to the limit to make it possible.

NEED FOR RISK MANAGEMENT

Perhaps in one sense, there is value in the distinctions between assessment and management. It is not the same as the distinction between science and policy. But if assessment is the art of determining the risk and management is the search for measures to reduce the risk, then I think we are today putting far too much emphasis on assessment and too little on management.

Through our single-minded concentration on assessment, we are entering a black and white world where a chemical is either a public menace or it is not. We seem to have lost the sense that man has been using hazardous materials since the dawn of time. Even a log on fire is a tremendous hazard, if you leave it in the wrong place. Through most of history we have been satisfied with fairly imprecise measurements of the risks we face and imperfect balances between risks and benefits. But through that whole history we have also worked mightily to reduce risks by applying new technology to old practice.

Perhaps the reason for this shift in emphasis from management to assessment is a simple measurement of the distance we have come. Life was for most of our history brutish and short. Risks came in gross quantities day in and day out. And better management could provide substantial benefits. Today, risks are measured in quantities of ten to the minus six (10^{-6}). Whether that is the reason for the change or not, I think it is time to reassert the emphasis on management. Knowing the risk is not the whole answer, but how to derive benefit from a chemical or an activity while managing risks at acceptable levels is the whole question.

REVIEW OF FEDERAL REGULATION: TSCA AND SDWA

TSCA

As chairman of the Toxic Substances and Environmental Oversight Subcommittee, reauthorization of TSCA is one of my high priorities for the coming year. The statute was passed in 1976 and has never been amended. Because implementation did not really get underway until 1980, we are only beginning to learn the lessons of TSCA.

I know your industry believes that TSCA is fine as it is, it does not need amendment. But I have decided to proceed as though the whole Act were open to question. We will analyze every section and every regulation promulgated by EPA under the authority granted to it. If we decide to reauthorize TSCA without amendment—and I think that an unlikely prospect—it should occur because Congress believes TSCA is working, not because Congress has too little time or inclination to look at it in detail.

TSCA is not only a gap filler. It should not function just to catch those problems not solved by some other environmental law. In fact, when you look at the other major statutes, they are not working well at all for toxic pollutants. As we go through this round of reauthorizations, we are presented in every case—whether it be Clean Air, Clean Water, Safe Drinking Water, or the Resource Conservation and Recovery Act (RCRA)—with proposals to toughen and tighten the regulation of toxic pollutants. When someone offers a list of chemicals, EPA is instructed to regulate those chemicals as an air or water pollutant or waste by a certain date. If EPA fails, a hammer falls on the industries that generate and use the chemicals. Whether the listing process will work is open to question.

But that is not how it should work. TSCA should function as the base for these other statutes. It should be used to develop information on hazard, exposure, and control. And that information should then be transferred to other media offices at EPA for the purpose of regulation. If TSCA were functioning as it might, Congress would feel no need to issue a list of its own.

We have already begun the hearing process for TSCA in the subcommittee. Additional hearings focusing on specific sections of the law will begin next year.

But there are items of apparent concern already. One is information. Too much of the information submitted to EPA under TSCA is stamped CBI, confidential business information. There are legitimate trade secrets, of course, but the cloak of secrecy is being thrown too broadly. EPA does not challenge industry claims to confidentiality because it lacks the resources to do so. But those with a legitimate and compelling need to know, including state health agencies and labor health specialists, should have access to the information. A strategy allowed under TSCA that needlessly denies information to the public cannot ultimately work to your benefit.

A second area of concern is the international coordination of risk assessment. The United States should support the international coordination of risk assessment through entities like the Organization for Economic Cooperation and Development and the International Programme on Chemical Safety. We should support their efforts for two reasons. It is cheaper and more efficient, since many chemicals are in the world marketplace, not just the U.S. marketplace. Secondly, public demand for protection from unreasonable risk is not unique to this country and we need to assure consumers all over the globe of safety, if we intend to keep our trade markets open.

SDWA

In addition to TSCA, the Environmental Oversight Subcommittee will reauthorize the Safe Drinking Water Act (SDWA) and the EPA research and development program. We will also devote considerable time to ground water as a resource asking whether groundwater protection requires a law all its own.

The problem with the Drinking Water law is quite different from the concerns about TSCA. The SDWA clearly does not work. The General Accounting Office conducted an audit of the Act in 1980 and found some 160,000 violations. How many other federal laws do you know that are violated 160,000 times a year and violated in many cases by other governments in our federal system?

I think SDWA was designed not to work. It appears Congress intended to force small water supply systems out of business by piling so many mandates on top of them, that you simply could not afford to be a consumer on a small water supply system. Water quality tests are still quite expensive. Few systems can employ experts to conduct the tests, yet the tests are required with great frequency under SDWA.

If the purpose of this Act was to encourage regional water supply systems, it simply has not worked. People chose to violate the federal law instead. The purpose of the Act—like all environmental and health statutes—should be to protect the public's health. If we want to do that in the case of drinking water, we need to offer assistance to these small systems rather than beat them about the ears. How can we make the tests cheaper? Can we make testing expertise widely available through circuit rider programs? What is a reasonable testing schedule for potential pollutants of various types?

Ground water, which is the drinking water supply for more than 50% of our population and which provides a vast resource for other uses, is not the direct subject of any law on the books. The Clean Water Act deals with discharges into surface waters. With the exception of one section on injection wells, the SDWA deals with water quality problems only after contamination has occurred. Superfund and the Solid Waste Statute—RCRA—are effective in protecting ground water only when you concentrate pollutants at one point and call it a dump or surface impoundment.

GAINING PUBLIC TRUST

Finally, what should your industry do about your problem? You do have a problem. The American public is afraid of you and your products. Let me quote from a recent article by William G. Simeral, your past board chairman. "In a sense we have a one-item agenda; all the major issues facing us flow from the fear of chemicals, their presumed toxicity and their potential impact on human health."

Let me cite another evidence of this fear. I was scheduled to appear at a recent event in New Jersey on acid rain. One of the panelists was Lou Harris, the public opinion pollster. Although I was unable to attend, I did have a chance to review Mr. Harris' written comments. They included a report on a recent poll conducted by his company to survey public attitudes on environmental issues, particularly acid rain and hazardous waste. What do you think he found when he asked the American public to name the principal cause of acid rain? According to the Lou Harris poll, a majority of 93 to 5 percent of the American public believes emissions from chemical plants are an important cause of acid rain. Sixty-four percent think that chemical plant emissions are a very important cause. Those represent higher margins than for any other source, including emissions from coal-fired utilities, which came in second, and emissions from automobiles and trucks, which came in last. As environmental issues go, the public has a great deal of information about acid rain. That they could so completely misunderstand the causes does not so much reflect the obscurity of the issue as it does the public's misperception of your industry.

What can you do about it? One step that I know many of you are already taking is to be aggressive in the effort to cleanup chemical dump sites. You have rightly understood that if you wait for the government to go through the motions of Superfund—and if in the process you give the impression that you are more concerned about your potential liability than about public health—then matters will become worse before they get better. Priority list or no, every community is concerned about the site or sites in their backyard. And it will be sometime and many dollars before we even get through the sites on the priority list. That you should commit your experts and your resources to addressing public concerns, not waiting for the government agency to flag every potential problem, seems to me the right thing to do.

Secondly, I think you need to encourage the development of effective tools and technologies for risk management. I know you do this already. But the control strategies you have developed have not been converted into tools of the government for regulation. Has anybody ever gone to EPA and volunteered a chemical for a SNUR—a Significant New Use Rule? I know many of your companies as a matter of routine policy engage in additional testing as your products move from one use to another or become more widely used in the marketplace. But has any company ever offered its internal policy to EPA as an appropriate rule?

Has anybody from your industry told EPA that Section 6 is underdeveloped and underutilized? That if more thought went into the management and control of exposure, less controversy would swirl around the assessment of risk?

Third is the matter of information, which I mentioned earlier. Turning information over to the government under the seal of confidentiality is not going to help your industry, because the public does not trust the government much either. That the government stands as a watchdog will not calm a public that thinks the dog is old and lazy and beset with fleas.

You need to flood the public record with information about your products—all the information you can generate all of the time. How is the CBI decision made at your company? Is it like the presumption of innocence? CBI unless proven otherwise? It should be the reverse. It should take a decision right at the top to deny the public—which remember is different from the government—access to information.

Finally, and this will be your most difficult task, you need to understand thoroughly public attitudes and the public thought process on risk and human health: the way we talk about it; the way it is communicated in the media; the kinds of government agencies that make pronouncements on it; and the other sources of information on which the public relies. This is not the science of chemistry, nor the province of the ad man. It is one of our most difficult public policy issues today. How shall we assess risk? How shall we decide what risks are acceptable? How shall we manage risks? And how can we make reasoned choices—choices that assure us the benefits of our age while protecting us from unreasonable risk—in a society that is open with a government responsive to people and to their concerns?

SUMMARY AND CONCLUSIONS

You need to be educators, to teach the public. Not just that chemicals are for better living, but how to think about risk. How to talk about it. How to choose in a risky society. Let me leave you with one more quote, from a book by Mary Douglas and Aaron Wildavsky called Risk And Culture. "Life's choices, after all, often come in bundles of goods and bads, which have to be taken as a whole. There is no sense in acting as if one can pick the eyes out of the potato of life, making entirely discrete choices when it comes all tied up, the bad mixed with the good."

CHAPTER 8

INDUSTRIAL VIEWPOINT AND CASE HISTORIES

Dr. Calvin J. Benning
Essex Chemical Corporation

INTRODUCTION

The Environmental Protection Agency (EPA) and industry have shared in the learning experience of implementing the Toxic Substance Control Act (TSCA). We both recognize that risk assessment and risk management have distinct differences and objectives, yet both are essential elements in the evaluation and control of potential hazards of chemicals.

The chemical industry conducts extensive testing of chemicals to determine their potential harm to both health and the environment. This testing process must, by its very nature, be a joint planning effort. First, it must ensure that scientifically sound procedures are followed. Second, data must be reliable and capable of being used to draw proper conclusions. And third, data must support proper regulatory decisions. Since the specific expertise of existing chemicals lies in the industry that knows its chemistry and the responsibility of implementing TSCA lies in the agency, it is natural that testing parameters be a joint decision process.

Difficulties have arisen in this process of implementing existing chemical programs. Three of these are:

- The natural process of defining goals and producing data is sometimes slow and arduous.
- There is a basic element of mistrust generated by the media between the environmentalists, the EPA, and industry.
- There is a lack of understanding of chemistry and the scientific process by both environmentalists and political parties.

These papers deal with this false perception that "industry is doing nothing." They also deal with product stewardship, risk management as a partnership, and specific case histories.

CHAPTER 9

A CASE HISTORY—PHTHALATES SECTION 4
TEST DATA USE IN REGULATORY CONTROL DECISIONS

Dr. James P. Mieure
Monsanto Company

INTRODUCTION

Phthalate esters are widely used as plasticizers. Numerous health and safety studies have been conducted on phthalates and several products are sanctioned by the Food and Drug Administration (FDA) for food packaging use. The Environmental Protection Agency (EPA) and other regulatory agencies have recently focused attention on phthalates, because of large production volumes, potential consumer exposure and positive rodent bioassay results for the most widely used phthalates.

The Chemical Manufacturers Association (CMA) Phthalate Esters Panel negotiated a voluntary testing program with EPA in 1981. The program consists of ecological and health effects testing covering the 14 most commercially significant products in the U.S. Tests are organized into phases, with shorter-term screening tests followed by longer term, more definitive tests based on the screening results. After preliminary phases are completed, the results are reported to EPA. Also, results are jointly evaluated by industry and EPA scientists at significant checkpoints in the program. Several disagreements on interpretation of results or specifics of test protocols have been resolved by discussions with the Agency.

The voluntary approach, in contrast to a mandated rule, incorporates the flexibility of scientific judgment to guide the assessment of completed studies and the planning of subsequent tests. Furthermore, test results will be available earlier under the voluntary program than would have been possible under a mandated rule.

The program is acquiring the data necessary to perform risk assessments on various phthalate esters. For ecological risk assessment the margin of safety concept (comparing concentrations which cause adverse effects to the concentrations in the environment) will likely be employed for aquatic organisms. A similar safety factor approach can be used to assess human safety for phthalates not suspected as potential carcinogens. If carcinogenicity is an issue, risk assessment models will likely be used. It is important when using models that all biological data be available to help ascertain which model best predicts the risk to man.

Once risk assessments are completed, the margin of safety or the risk estimate must be compared to values acceptable to society. Values which seem appropriate from previous regulatory experience are presented.

COMMERCIAL USE OF PHTHALATE ESTERS

Phthalate esters are a class of chemicals formed by reacting two moles of alcohol with one mole of phthalic anhydride. Differences in molecular weight or structure of the alcohols lead to different properties in the final ester. Phthalate esters have been widely used as plasticizers in U.S. commerce for about 40 years with no apparent harm to humans or the environment. During this 40-year period, several hundred health and environmental safety tests have been conducted on phthalates by producers and by government and academic institutions. Based on these test results, several phthalates are sanctioned by FDA as indirect food additives for use in food packaging.

REGULATORY FOCUS

Recent regulatory interest in phthalates can be traced to 1976 when, under a court decree, EPA named phthalate esters as one of 65 chemicals or categories designated as Priority Pollutants under the Clean Water Act. Further definition limited the category to six phthalates of interest. In 1977, Alkyl Phthalates were nominated by the Interagency Testing Committee (ITC) for priority attention for potential Test Rule development by EPA under the Toxic Substances Control Act (TSCA). This nomination focused on ecological effects and was justified partially on production volumes, use in consumer articles, and reports (later shown to be unsubstantiated) of widespread presence in the nation's waterways.

In 1980, interest in human health issues with phthalates was heightened when the National Toxicology Program (NTP) published draft reports indicating that di(2-ethylhexyl) phthalate and di(2-ethylhexyl) adipate caused liver tumors in lifetime rodent feeding studies. Also in 1980, an alkyl aryl phthalate, butyl benzyl phthalate, was nominated by ITC for TSCA Test Rule consideration. EPA's Test Rules Development Branch rejected the manufacturer's request to have this substance's large existing data base evaluated independently by the Agency and decided to include butyl benzyl phthalate in the alkyl phthalate program. In 1981, the Interagency Regulatory Liaison Group (IRLG) was convened to promote health-related communications. During this time data assessment was underway between scientists from industry and EPA. More about that in the next section.

The first actual regulatory decision by EPA on a phthalate occurred in 1982 when EPA, acting in response to a citizen TSCA 4(f) petition as well as staff evaluations, published a statement that di(2-ethylhexyl) phthalate (DEHP) did not meet the 4(f) criteria to be considered as a "substantial risk" substance.[1/]

In contrast to this decision by EPA, the Consumer Product Safety Commission (CPSC) authorized an assessment program on uses of phthalate esters in consumer products as one of their ten fiscal year 1983 priority projects. In September, 1983, the Commissioners voted to convene a Chronic Hazard Advisory Panel to consider the risk posed by children's exposure to DEHP in pacifiers, teething rings, toys and other articles.

1/ U.S. Environmental Protection Agency Press Advisory, "Review of Data Available to the Administrator Concerning Formaldehyde and Di(2-ethylhexyl) Phthalate (DEHP)," February 12, 1982.

COOPERATIVE INDUSTRY/EPA EFFORTS

In 1980, the existing CMA Phthalate Esters Panel was restructured and turned its attention from sponsoring studies on ecological properties of phthalates to a broader-based health and environmental effects testing program. More than 30 companies committed resources to a joint program aimed at advancing the current state of knowledge on health and environmental safety of phthalates. Industry scientists and toxicologists met several times to share unpublished data and to develop an integrated perspective on known properties of phthalates. Throughout 1981, a series of meetings were held with scientists and management of EPA's Test Rules Development Branch, and gradually a consensus emerged on the current state of knowledge on phthalate ester toxicology and the data gaps which still needed to be filled. Further discussions, including resolution of details of testing protocols, led to EPA's acceptance of the CMA Panel's voluntary testing program in late 1981. This was the first industry voluntary testing program to be endorsed by the Agency in lieu of a mandatory testing rule under Section 4 of TSCA. Testing at contract laboratories actually began prior to EPA's final acceptance of the program. Because of the cooperative industry/EPA data assessment and the prompt initiation of testing, new data on phthalates will be available earlier under this negotiated program than under a mandated test rule.

SUMMARY OF VOLUNTARY TEST PROGRAM

The program sponsored by the industry consortium consists of tests to assess the ecological and human safety of the 14 phthalate ester products manufactured in largest volume in the U.S. All tests are subject to rigid quality assurance audits by the Panel and by EPA. The ecological portion of this program is comprised of tests to determine the acute and chronic toxicity of representative products to various aquatic organisms.[2/] Another facet of the ecological

---

2/ River and lake sediment has been shown to be the environmental compartment to which phthalates would migrate should there be environmental release.

tests is a determination of the degradation rates of representative phthalates by the most likely breakdown process, which for most phthalates is biodegradation by microorganisms.

The health effects portion of the program has several elements. One portion consists of short-term mutagenicity and biological tests to help identify any compounds having significant oncogenic potential that might warrant further testing. DEHP is used as a reference compound for these studies. Another portion of the program is metabolism and pharmacokinetic studies to better understand the mechanism of action for DEHP in rodents and the suitability of rodents for assessing phthalate risk to humans.

The CMA Phthalate Panel's program is organized into phases, with longer-term, more definitive tests in the latter phases preceded by shorter-term screening tests. Data from the screening phases are used to establish the conditions (especially dosages) and to select compounds for conducting the more definitive tests.

With a complex testing program as described above, it was not practical a priori to establish a rigid set of decision factors to provide guidance for all scenarios which might develop as the testing progressed. It was mutually agreed that industry and Agency scientists would evaluate test results at significant checkpoints and jointly determine the need for and details of future tests. The program will terminate when a sufficient data base has been generated to address the concerns of EPA and industry.

The preliminary phases are completed and the results reported to EPA's Test Rules Development Branch. EPA personnel have been apprised of key developments of each phase and have been consulted at the significant checkpoints in the program. In several instances, Agency and industry scientists have disagreed on specific elements as the program developed. In each case, these differences were resolved by appropriate discussions with the Agency. Necessary protocol and scope changes have been dealt with much more effectively by these discussions than by attempting to incorporate provisions for needed technical changes into formal rule making. Thus, the voluntary program is proceeding at a pace equivalent to that of a mandated Test Rule and will actually be completed earlier. This earlier completion will be accomplished because many of the steps involved in promulgating a Test Rule were not necessary and the program was initiated as soon as conceptual agreement was announced.

RISK ASSESSMENT

The risk assessment process requires the collection of toxicity information about a substance. To be useful, toxicity data must be combined with available exposure information to develop a perspective on the degree of risk posed by the substance.

For ecological risk assessment, this process is relatively straightforward. Toxicity studies can be conducted in the laboratory directly on organisms representative of the species in need of protection. In the CMA Panel's program, this includes freshwater and saltwater fish and invertebrates. These studies demonstrate the levels below which no adverse effects are observed. The chronic no-effect levels can then be compared to the levels predicted or measured in environmental surface waters. This comparison can be expressed as a margin of safety. The numerical value of this margin of safety which is judged to be adequate to protect aquatic organisms will depend somewhat on the degree of certainty of the exposure estimates or measurements. A margin of safety of ten might be acceptable based on measured environmental concentrations while 100 might be acceptable based on predicted exposure.

For human risk assessment the toxicity data are normally collected on animals. Thus extrapolation is necessary from animal models to man. If the effects observed in animals do not include cancer, the risk assessment process is similar to that described above for ecological organisms. The highest dose level at which no adverse effects were observed in animals is divided by a safety factor (usually 100). This potential exposure level is often termed the "acceptable daily intake" and is generally considered to be a safe level of exposure for humans.[3/] This approach has been used successfully for many years for food additives in the United States and most other Western countries. Of course, to be generally applicable to situations not restricted to oral intake, all routes of exposure would have to be included when determining the human daily intake.

3/ J. Rodricks and M. R. Taylor, "Application of Risk Assessment to Food Safety Decision Making," Regulatory Toxicology and Pharmacology, III, No. 3, September, 1983 (in press).

For substances which cause cancer in animals, the risk assessment process is more complicated. Generally, one or more mathematical models will be used to provide quantitative estimates of risk to humans. Use of any model requires a selection of the proper measure of dose (i.e., target site versus administered) and an examination of the mechanism of action. An assumption about interspecies extrapolation is necessary, as well as an estimate of human exposure. All available biological information should be considered as part of the assessment.

Sufficient information from animal studies and simulated exposure tests is available to make a preliminary risk estimate on DEHP, a widely used phthalate plasticizer. Testing in progress will provide a firmer foundation for an assessment in the future. While preferring to await more complete data, the CMA Phthalates Panel has commissioned a preliminary risk assessment on certain uses of DEHP encompassing the available mechanism and exposure information.

The preliminary work has focused on modeling of rodent carcinogenicity data and calculating dose coefficients. One widely used model which has substantial basis in cancer causation theory is the "multistage" model. Using this model without applying any pharmacological or interspecies extrapolation correction factors which would mitigate the risk, one can derive a conservative risk estimate for various potential exposure concentrations. For example, if one were to select a dose of one μg/kg/day (which is within the range of recent CPSC estimates), preliminary calculations indicate the risk of an individual contracting liver cancer due to exposure to DEHP is less than one in a million. Use of other risk assessment models which are compatible with available information on DEHP's mechanism of action would lead to even lower risk estimates. Risks this low are not normally considered relevant for regulatory consideration.

REGULATORY DECISIONS

In the discussion of risk assessment, quantitative values were mentioned as acceptable for demonstrating the safety of a substance. In summary, these values were a "margin of safety" of ten to 100 for protection of aquatic organisms and a human "acceptable daily intake" a hundred-fold below animal effect levels for substances lacking evidence of animal carcinogenicity. In each case, the "margin of safety" or "acceptable daily intake" implies a regulatory judgment that risks below these values are acceptable to society. For quantitative risk assessments of animal carcinogens, similar judgments have been made. FDA, OSHA, and EPA have all stated that substances having risks below one in a million ought not be subjected to regulation.

SUMMARY AND CONCLUSIONS

The CMA Phthalates Panel believes these values represent reasonable guidelines for regulatory decisions on phthalate esters. Phthalates having risks below these levels should not be regulated. Phthalates with risks above these levels may need to be subjected to some form of risk management, provided there is a significant population at risk. Actual application of these principles to regulatory practice awaits the completion of our test program.

BIBLIOGRAPHY

Environmental Protection Agency Press Advisory. "Review of Data Available to the Administrator Concerning Formaldehyde and Di(2-ethylhexyl) Phthalate (DEHP)." February 12, 1982.

Rodricks, J. and Taylor, M. R. Regulatory Toxicology and Pharmacology, III 3, No. 3, September, 1983 (in press).

CHAPTER 10

VINYL CHLORIDE AND TSCA

John T. Barr
Air Products and Chemicals, Inc.[1/]

INTRODUCTION

The well-known regulatory history of vinyl chloride and its role as a bellwether of current regulatory philosophy makes it a useful paradigm for examining the relationship of existing laws and the Toxic Substances Control Act (TSCA) for control of chronic hazards.

To this end, we will first review some of the highlights of its regulatory history, and then engage in some speculation as to the response these events might elicit today under TSCA.

INDUSTRIAL AND COMMERCIAL USE OF VINYL CHLORIDE

Vinyl chloride became of industrial importance about fifty years ago, approximately a hundred years after its discovery, when Semons discovered that its polymer could be converted into useful articles by plastization with phthalate esters. Commercial development began first in Europe and then in this country in the late

---

1/ Air Products and Chemicals, Inc., 1983.

thirties, largely using existing rubber processing equipment, for it was rubber which it initially replaced in the market. For the same reason, the use of polyvinyl chloride (PVC) was sequestered by the government during the war years, and it was not until the early fifties that widespread consumer applications developed. PVC is now a mature product, and its growth rate falls in step with the Gross National Product. Presently, about six billion pounds are used annually in this country, and about four times that in the world.

Some of the broader toxicological attributes of vinyl chloride (VC) were recognized in the thirties. It was known to be an anesthetic, but problems with cardiac arrythmia prevented its use in that application.[2/] As pathological techniques improved, industry scientists recommended in the early sixties that exposure be limited to 50 ppm, because of temporary liver enlargement in animals at that level,[3/] but the American Conference of Governmental and Industrial Hygienists considered this overly conservative and accepted instead the 500 ppm recommendation of Harvard scientists.[4/] This was the value adopted by the Occupational Safety and Health Administration (OSHA) in its formative days.

Also in the early sixties, the European industry recognized among its workers a disease termed acroosteolysis, AOL, which is a degenerative disease of the bone tufts, particularly in the fingers, that is accompanied by Reynaud's phenomenon.[5/] An extensive epi-

---

2/ W. F. von Oettigin, "The Halogenated Hydrocarbons, Their Toxicity and Potential Dangers," Public Health Service Publication No. 414 (Washington, D.C.: U.S. Department of Health, Education and Welfare, 1955).

3/ T. R. Torkelson, F. Oyers, and V. K. Rowe, "The Toxicity of VC as Determined by Repeated Exposures of Laboratory Animals," American Industrial Hygiene Association Journal, XXII (1961), p, 354.

4/ American Conference of Governmental and Industrial Hygienists, "Documentation of the Threshold Limit Value, 1963" (Cincinnati, OH, 1963).

5/ S. Suciu, J. Drejman, and M. Valaskai, "Study of Diseases Caused by Vinyl Chloride," Medical Intern., XV (1963), p. 967.

demiological survey here and in Europe found about a hundred possible cases which were associated closely with manual cleaning of reactor walls between polymerization batches, but neither the precise etiological agent nor the disease mechanism was identified.[6/]

An attempt was made to reproduce this disease in rats by the medical department of one of the European producers. An exact duplication of the human disease was not seen, but many of the rats developed tumors at numerous sites. The reporting of this finding by Viola[7/] in 1970 evoked little interest in the regulatory community, possibly because of the very high doses used, several thousand ppm, which were frankly toxic to the animals, and the fact that the tumors were largely metastatic from the Zymbal gland, an organ not present in humans.

Nevertheless, both the European and domestic producers formed consortia to perform bioassays at lower concentrations and also began epidemiological surveys of their employees.

Preliminary results of the European bioassay became available first in early 1973, and showed tumor development at much lower concentrations in organs which do have human counterparts. This result was transmitted to regulatory officials that summer, and industry screening of employee records was intensified.[8/] This resulted in the recognition that winter by an industry medical director of a cluster of three rare liver tumors termed angiosarcoma, ASL, in the employees of one facility.[9/] The reporting of this fact to

---

6/ W. A. Cook, et al., "Industrial Hygiene Evaluation of Thermal Degradation Products from PVC Fetus in Meat-wrapping Operations," Arch. Environ. Health, XXII (1971), p. 74. Also, B. D. Diman, et al., "Occupational Acroosteolysis I, An Epidemiological Study," ibid., p. 61.

7/ P. L. Viola, "Pathology of Vinyl Chloride," Medicina del Lavoro, LXI (1970), p. 174.

8/ A. W. Barnes, "ICI Ends Its Silence on Vinyl Chloride," Chemical Engineering News, (July 8, 1974), p. 21.

9/ J. L. Creech and M. N. Johnson, "Angiosarcoma in Workers Exposed to Vinyl Chloride as Predicted for Studies in Rats," Journal of Occupational Medicine, XVI (1974), p. 150.

government officials led to the current regulatory status of vinyl chloride.

It also led to a virtual explosion of research on the chronic toxicity of VC. The body of scientific literature on the oncogenicity of vinyl chloride is as large as that for any other substance. It is recognized that VC is a classical procarcinogen. Metabolism by the mixed function oxidase in the liver converts it to the ultimate carcinogen, an epoxide. Detoxification of this intermediate by the sulfhydryl group of glutathione or other proteins removes the toxic potential.[10/] Both of those mechanisms are saturable.[11/] An overload of the metabolic step assures that the vinyl chloride will pass through the liver and some will be metabolized in other organs. An overload of the detoxification step allows escape of the toxicant into the sinusoidal passages of the liver where interaction with the chromosomal protein causes ASL to develop. An overload of both mechanisms can lead to tumor development outside of the liver, as is seen in mice and rats at very high doses. Despite the large data base, however, information on the precise mechanism of these various steps still is lacking. We do not even understand why some persons respond with AOL and some with ASL, but none with both diseases.

REGULATORY STANDARDS

OSHA proceeded promptly in early 1974 to set an emergency temporary limit of 50 ppm for worker exposure, and later that year reduced the limit to one ppm, the current figure. Industry was given a grace period during which respirators could be used to meet this requirement, but now that level must be met by engineering practices.[12/]

10/ W. K. Lelbach and H. J. Marsteller, "Advance in Internal Medicine and Pediatrics " Springer-Verlag, XLVII (New York, 1981).

11/ R. Hefner, P. Watanabe, and P. Gehring, "Percutaneous Absorption of Vinyl Chloride Gas in Rhesus Monkey," Toxicology and Applied Pharmacology, XXXIV (1975), p. 529.

12/ OSHA Standard for Vinyl Chloride, 29 CFR 1910.1017.

The Environmental Protection Agency (EPA) promulgated a combined engineering and works practice standard in 1976 which has resulted in ambient concentrations in the fractional ppb range near producing or using facilities.[13/]

In the meanwhile, the Food and Drug Administration (FDA) and Consumer Product Safety Commission (CPSC) established prohibitions on the use of VC in aerosol or other consumer applications, a practice which had been discontinued in 1973. The Bureau of Alcohol, Tobacco and Firearms of the Treasury Department (BATF) had already banned the use of PVC liquor bottles in 1973 because of concern for taste effects from migration of residual VC into the contents. In 1975 the FDA proposed revocation of the generally regarded as safe (GRAS) status of rigid PVC packaging under the Delaney clause, also because of migration concerns, but that proposal never has been promulgated, and the FDA has stated that it is considering withdrawal of the proposal and recommending to BATF the reauthorization of plastic liquor bottles in light of the current very low residual monomer levels in fabricated PVC articles.

Other regulations have followed as new statutes and rules have come into play. The Department of Transportation (DOT) and the Coast Guard regulate the transportation of VC, of course, and VC is listed as a priority pollutant and hazardous waste under various water and solid waste rules, and has a reportable quantity of one pound under Superfund.

Did the existing laws operate satisfactorily at the time of discovery of the chronic hazards of VC? It appears that they did. A leading medical authority who was deeply involved in the worker health evaluation in 1974 has termed VC a "success story." Reevaluation of the risk to employees under the one ppm standard by a conservative nonthreshold extrapolation method[14/] yields a lifetime estimate of less than 10^{-8}, a risk level which is not thought to be of concern. The comparable risk estimate for the general populace is

---

13/ EPA Standard for Vinyl Chloride, 40 CFR 61.60.

14/ P. J. Gehring, P. G. Watanabe, and C. N. Park, "Risk of Angiosarcoma in Workers Exposed to Vinyl Chloride as Predicted for Studies in Rats," Toxicology and Applied Pharmacology," XLIX (1979), p. 15.

several orders of magnitude lower. EPA has stated on several occasions that it believes that vinyl chloride is regulated adequately.

RISK ASSESSMENT

Risk assessment has been a popular avocation among those interested in VC, and more than a dozen have been performed.[15/] These can be divided generally into two classes: those which rely solely on animal data; and those which attempt to incorporate the human experience.

Those in the first class yield similar results, and show the normal spread of estimates from the various mathematical models in common use. These range from 1,500 to 10^{-5} ppb for a lifetime risk of 10^{-6}, or eight orders of magnitude. It is necessary to eliminate the high-dose data points, that is, those over 2,500 ppm from the Maltoni data[16/] in order to get reasonable fits to most models, because these doses show broad systemic toxicity. The lower doses, 500 ppm and below, as a group fall into a general pattern on a log-probit plot, but individual two or three dose experiments show tremendous differences in slope when plotted separately. The popular multihit model predicts a lifetime risk of 10^{-6} at fractional ppb levels.

The human factor was accounted for in two ways. The EPA used some preliminary employee epidemiological data to confirm its animal-based extrapolation.[17/] Unfortunately, the human data were

---

15/ J. T. Barr, "Risk Assessment for Vinyl Chloride in Perspective," (Paper 82-9.2 presented at the 75th Annual Meeting of the Air Pollution Control Association, New Orleans, LA, 1982), Lines 20-25.

16/ C. Maltoni, et al., "Vinyl Chloride Carcinogenicity Bioassays (BT Project)," (Paper presented at "Le Club de Cancerogenese Chemique," Institute Curie, Paris, November 10, 1979).

17/ A. M. Kusmack and R. E. McGoughy, "Quantitative Risk Assessment for Community Exposure to Vinyl Chloride," (Washington, D.C.: U.S. Environmental Protection Agency, December 5, 1975).

selected from those locations known to have ASL cases, while other facility data were omitted. They also were in error on the past exposures by more than an order of magnitude. This resulted in an estimate of 20 cases per year from the estimated 1974 ambient concentrations for the population within five miles of production and processing facilities.

The EPA seldom bothers to check its estimates against available data, so it sometimes comes up with results such as that made for arsenic a few years ago that would have predicted 18 million cases of skin cancer a year in this country if it had been applied to Agency data on the average arsenic concentrations in drinking water. Similarly, a survey of all known ASL cases in this country for the ten years before 1974 showed no cases associated with residency near such plants,[18/] rather than the 200 predicted cases. It is reasonable to assume that if any cases had developed since that time, the publicity associated with it would have brought them to light. Thus we have 110 million-person years of negative history for nearby residents. This places an upper limit on risk of less than 10^{-7} per ppm-yr.

Two studies applied pharmacokinetics in an attempt to obtain relevant human data. Gehring and coworkers estimated a lifetime risk of 10^{-8} at one ppm from the probit model, based on a biotransformation of rat data. The unconstrained linear model predicted no risk at less than 99 ppm.[19/]

Anderson, Hoel and Kaplan carried this procedure one step further, and applied it to bound metabolic products, rather than to the total amount metabolized. Their results gave a lifetime risk of 10^{-7} at less than one ppm, with the probit model, or at less than two ppm with the linearized multistep model.[20/]

18/ H. Popper, et al., "Development of Hepatic Angiosarcoma in Man Induced by Vinyl Chloride, Thorotrast, and Arsenic," American Journal of Pathology, XCII (1978), p. 349.

19/ P. J. Gehring, P. G. Watanabe, and C. N. Park, Toxicology and Applied Pharmacology, XLIX (1979), p. 15.

20/ M. W. Anderson, D. G. Hoel, and N. L. Kaplan, "A General Scheme for the Incorporation of Pharmacokinetics in Low-dose Risk Estimation for Chemical Carcinogens. ibid., LV (1980), p. 154.

Thus we see that risk is in the eye of the estimator, but it is clear that estimates incorporating human data reflect the human experience for VC far better than do the direct application of animal data.

There was understandable uncertainty on the part of both the regulators and industry in 1974. This was the first commodity chemical to be regulated under the relatively new statutory situation as the result of new information. Nevertheless, both the regulatory agencies and industry acted promptly to reduce exposures and emissions to an acceptable level.

The current count of occupational ASL cases is about 100 worldwide, with 30 of these in this country.[21/] All these cases had their first exposure in 1964 or earlier, and there appears to be room for optimism that the steps taken in the mid-sixties because of the AOL information will have prevented any significant number of cases developing from exposures commencing after that date. Certainly it is reasonable to expect that there have been no new cases initiated after the early seventies.

Had TSCA been in place in the mid-sixties, would it have made any difference in the course of events? It appears unlikely that it would. Certainly the AOL discovery would have resulted in a series of 8(e) notices to TSCA. The probable outcome of that would have been either a recommendation from the Interagency Testing Committee (ITC) for more tests, or a Section 4 testing requirement. It is possible that, because of its commercial importance, VC could have been placed on the ITC list before the AOL data became available. Additional data could have been called for under Sections 8(a) and (d). The result of all this most likely would have been a negotiated testing rule, under which industry would have initiated a series of studies which would have culminated in a bioassay, and the carcinogenicity of VC would have been discovered in due time. Yet, this is precisely what did happen in the absence of TSCA, except that the preliminaries were omitted, and the bioassay was performed concurrently with the screening tests. Thus it is possible that the

---

21/ J. Stafford, personal communication, Liver Angiosarcoma Cases, April 15, 1983.

final critical data were obtained earlier than would have occurred under present conditions.

Bear in mind that most of today's powerful testing methods were not available twenty years ago. That fact would not have been changed by legislative fiat, and any decision made at that time had to be made in light of the available knowledge.

If the data of Viola suddenly became available today instead, would there be any significant difference in the outcome, or the timing of that outcome? Probably so, but only because of the vastly more powerful scientific tools which we have available to us now. Neither the speed of agency motion nor the rate at which industrial facilities can be built or modified has increased. If anything, the latter has slowed, given the multiplicity of permits and approvals now required. Overall, it is possible that if today we knew nothing more about VC than was known in 1970, we would arrive at a regulated state a few months earlier than was achieved in 1974, but scientific progress, and not legislative or regulatory advancement, should get the credit.

What if VC were to become a new product today? Would it run the same course in which it would be 40 years before there was full recognition of its chronic potential? Certainly not. Again, however, the reason is due more to scientific progress rather than statutory development.

One change might be apparent. If VC were the subject of a Premanufacture Notification (PMN) today, rather than being the model to which all other aliphatic olefins are compared for structure-activity analysis, it would be judged by the others in its family. This comparison would be less dogmatic than the reverse is now. Ethylene and vinylidene chloride are not animal carcinogens; the relevance to humans of the carcinogenicity of high doses of trichloroethylene (TCE) is equivocal and controversial; and vinyl acetate has only a preliminary "non-negative" report. Thus, this class of substances would have lost its leader for structure activity comparison, and a decision as to the need for further testing from that analysis would not be clear-cut, based on analogous compounds.

Neither would a full minimum premanufacture data (MPD) set be of any great assistance. VC responds poorly to the classical in-vitro tests, and only recently has it become possible to obtain reproducible positive results in many of these. If the position were taken that any positive result triggers further testing, then we would be left exactly where we were in the late sixties, recognizing the need for a bioassay.

One other point should be considered before the requirements of Section 9 of TSCA to give primacy to existing statutes is ignored. Section 2 of TSCA requires the consideration of economic factors in actions taken under TSCA, while the Occupational Safety and Health Act and the Clean Air Act Section 112 do not. In fact, at the time that VC was being regulated, these two statutes were being interpreted as forbidding economic consideration. Had TSCA been the regulatory vehicle of 1974, it is unlikely that the final regulations could have been as strict as they actually became, because of this factor.

It is difficult to separate cleanly the compliance costs for the OSHA and EPA rules on VC because of the overlapping time periods. The best estimate for OSHA costs made by the industry in a presentation to the Presidential Task Force on Regulatory Relief was something over $200 million of capital, with over $25 million per year of annual costs. The EPA has reported to Congress[22/] that compliance with its rule has cost about $100 million per year since 1978. Thus, about $900 million has been expended thus far. Various published costs per life saved have ranged from $4 to $200 million for the OSHA standard, depending on which exposure starting point was used. The more costly EPA standard appears not to have prevented any cases of ASL, based on epidemiology, and thus has an infinite cost.[23/] This suggests that the proponents of strong TSCA activity for existing chemicals should reconsider their position on Section 9 if their long-term goal is more stringent regulations.

There seems to be no sure method of preventing some surprises in toxicology. Improved surveillance and diagnostic methods assure that we will know more about chronic effects in the future than we do now. New substances simply cannot be subjected to full-scale

---

22/ Environmental Protection Agency, "The Cost of Clean Air and Clean Water," (Annual Report to Congress, Senate Document 96-38, December, 1979).

23/ J. T. Barr, "Risk Assessment for Vinyl Chloride in Perspective."

testing before they show strong commercial promise. If there were no other reasons for this, the limitation on test facilities dictates that we direct our immediate effort toward substances of major import, and this is being done at capacity. It is proper that we concentrate our efforts on present exposures. Fortunately, recent medical advances help us to recognize these surprises earlier, and to minimize their impact.

Existing statutes, that is, non-TSCA derived regulations, appear to be able to regulate existing substances adequately. The history of vinyl chloride bears this out. The principal value of the TSCA derived activities seems to be in the gathering of surveillance data on these substances to assure that the relevant data are made available to the proper agencies, and in future oversight of new substances as they develop into commercial items.

SUMMARY AND CONCLUSIONS

The reexamination of the hazards of all existing chemicals is an overwhelming task for which EPA has no special expertise.[24/] Just the establishment of priorities for such a reexamination is beyond the present capacity of the Agency.[25/] The Agency has recognized some of the problems which it faces, and the recent "TSCA Priorities and Progress, 1983" report discusses a much more sharply delineated existing chemicals program. Even here, however, the division between TSCA and existing statutes is not defined clearly. Further, the Agency appears to be entering the realm of risk management through its Advisory and Chemical Hazard Information Profile (CHIPs) series. One unfortunate result of this is the generation of another tainted list of substances which becomes an invitation for pressure to regulate. The temptation to prepare these

---

24/ National Research Council, "Regulating Pesticides," (Washington, D.C.: Environmental Studies Board, Committee on National Resources, 1980).

25/ J. T. Barr, "Establishing Regulatory Priorities," Toxic Substances Journal, IV (1983), p. 290.

"little lists" for the executioners apparently is too great to be resisted,[26/] as Lester Lave pointed out recently.

We believe that EPA can best obey its statutory mandate by developing a more efficient system for establishing priorities, and by implementing more effectively its Section 9 procedures.

BIBLIOGRAPHY

American Conference of Governmental and Industrial Hygienists. "Documentation of the Threshold Limit Value, 1963." Cincinnati, OH, 1963.

Anderson, M. W.; Hoel, D. G.; and Kaplan, N. L. "A General Scheme for the Incorporation of Pharmacokinetics in Low-dose Risk Estimation for Chemical Carcinogens." Toxicology and Applied Pharmacology," Vol. LV (1980), 154.

Barnes, A. W. "ICI Ends its Silence on Vinyl Chloride." Chemical Engineering News, (July 8, 1974), 21.

Barr, J. T. "Risk Assessment for Vinyl Chloride in Perspective." Paper 82-9.2, 75th Annual Meeting of the Air Pollution Control Association, New Orleans, LA (1982).

Barr, J. T. "Establishing Regulatory Priorities." Toxic Substance Journal, Vol. IV (1983), 290.

Cook, W. A. "Industrial Hygiene Evaluation of Thermal Degradation Products from PVC Fetus in Meat-wrapping Operations." Arch. Environ. Health, Vol. XXII (1971), 74.

Creech, J. L., and Johnson, M. N. "Angiosarcoma of Liver in Manufacture of PVC." Journal of Occupational Medicine XVI (1974), 150.

26/ Lester Lave, "The High Cost of Regulating Low Risks," Wall Street Journal, August 19, 1983.

Diman, B. D., et al. "Occupational Acroosteolysis I, An Epidemiological Study." Arch. Environ. Health, Vol. XXII (1971), 61.

Environmental Protection Agency. "The Cost of Clean Air and Clean Water." Annual Report to Congress, Senate Document 96-38, December, 1979.

Gehring, P. J., Watanabe, P. G., and Park, C. N. "Risk of Angiosarcoma in Workers Exposed to Vinyl Chloride as Predicted for Studies in Rats." Toxicology and Applied Pharmacology, Vol. XLIX (1979), 15.

Hefner, R.; Watanabe, P.; and Gehring, P. "Percutaneous Absorption of Vinyl Chloride Gas in Rhesus Monkey." Toxicology and Applied Pharmacology, Vol. XXXIV (1975), 529.

Kusmack, A. M., and McGoughy, R. E. "Quantitative Risk Assessment for Community Exposure to Vinyl Chloride." U.S. Environmental Protection Agency, Washington, D.C.; December 5, 1975.

Lave, L. "The High Cost of Regulating Low Risks." Wall Street Journal, August 19, 1983.

Lelbach, W. K., and Marsteller, H. J. "Advance in Internal Medicine and Pediatrics." New York: Springer-Verlag, Vol. XLVII (1981).

Maltoni, C., et al., "Vinyl Chloride Carcinogenicity Bioassays (BT Project)." Paper presented at "Le Club de Cancerogenese Chemique," Institute Curie, Paris, November 10, 1979.

National Research Council. "Regulating Pesticides." Environmental Studies Board, Committee on National Resources, Washington, D.C., 1980.

Popper, H., et al. "Development of Hepatic Angiosarcoma in Man Induced by Vinyl Chloride, Thorotrast, and Arsenic." American Journal of Pathology, Vol. XCII (1978), 349.

Stafford, J. Personal communication, Liver Angiosarcoma Cases, April 15, 1983.

Suciu, S.; Drejman, J.; and Valaskai, M. "Study of Diseases Caused by Vinyl Chloride." Medical Intern, Vol. XV (1963), 967.

Torkelson, T. R.; Oyers, F.; and Rowe, V. K. "The Toxicity of VC as Determined by Repeated Exposures of Laboratory Animals." American Industrial Hygiene Association Journal, Vol. XXII (1961), 354.

Viola, P. L. "Pathology of Vinyl Chloride." Medicina del Lavoro, Vol. LXI (1970), 174.

Von Oettigin, W. F. "The Halogenated Hydrocarbons, their Toxicity and Potential Dangers." Public Health Service Publication No. 414, U.S. Department of Health, Education and Welfare, Washington, D.C., 1955.

CHAPTER 11

INDUSTRY'S PERSPECTIVE ON HOW INDUSTRY/GOVERNMENT RISK MANAGEMENT PARTNERSHIP CAN WORK

Dr. John D. Behun
Mobil Oil Corporation

INTRODUCTION

I am reminded of a conclusion reached early in my college education that proved to be a truism throughout my life—that is, it is strange how much you must know before you realize how little you really do know.

The examples described by the preceding authors illustrate that both government and industry have a keen interest in and are making sincere efforts at risk assessment and risk management. It is apparent from these and many other examples that could be cited that we all have much to learn in this area. The purpose of this seminar is to bring us all up higher on the learning curve. In this presentation, we will analyze how this might best be accomplished.

RISK MANAGEMENT OBJECTIVES

The common goal for both industry and the Environmental Protection Agency (EPA) with respect to risk management is protection against unreasonable risk. Industry has a major role in the

RISK MANAGEMENT

GOAL: No unreasonable Risk

A Major Responsibility of Each:

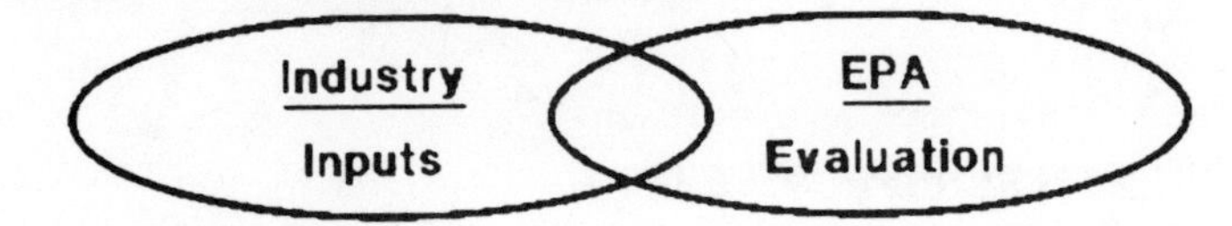

KEY ELEMENTS:

- Quantity of Data
- Quality of Data
- Separate RA From RM
- Peer Review
- Dialoque Throughout

Figure 1.

EPA EXISTING CHEMICALS PROGRAM

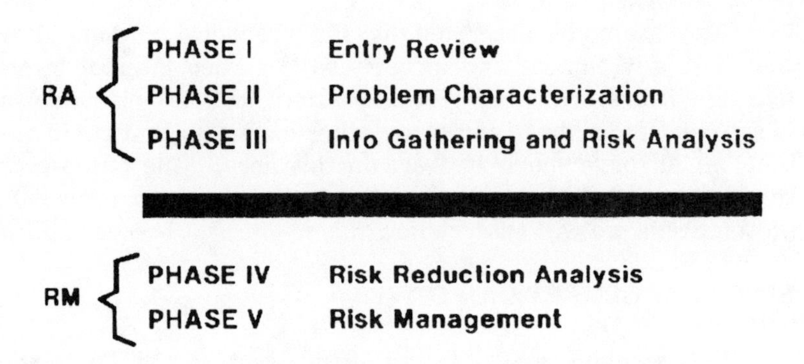

Figure 2.

process by providing EPA with input on up-to-date real life experience. One of EPA's principle duties should be a thorough evaluation of all available information. These responsibilities are interdependent, so we show them as overlapping (Figure 1).

Going a step further in our analysis, let us consider some of the key elements of the overall risk management process. All too often we find data inadequate both in quantity and quality. For many chemicals we seldom have more than sparse acute toxicology data. Also, the quality of data extracted from published literature often is in doubt because of obsolete test standards and practices.

Another key element is separation of risk assessment from risk management—a distinction with almost universal support. Risk assessment represents the science aspect of the process which then is incorporated in risk management along with social, economic, and policy considerations. However, since risk assessment is still far from a mature science and subject to a wide latitude of judgment, peer review by both industry and government scientists also is essential. Finally, meaningful dialogue must be maintained throughout to assure practical and workable risk management decisions that will not create confrontation or litigation. Although not shown here in the overlapping responsibilities, the dialogue also must include public interest groups with the technical expertise and ability to contribute constructively.

EPA'S EXISTING CHEMICALS PROGRAM

EPA's program for evaluating and controlling existing chemicals has been defined in terms of five distinct phases (Figure 2). The first three constitute steps leading up to the risk analysis. The last two phases are intended to arrive at the appropriate risk management decision. We will examine each of the five phases briefly from an industry perspective.

Entry Review

Phase I, Entry Review, is essentially a screening stage to determine those candidates which should be given priority consideration (Figure 3). The candidates are substances emerging from the Toxic Substances Control Act (TSCA) Section 4 program, TSCA

PHASE I - ENTRY REVIEW

- Screen Candidates
- Prepare Chemical Hazard Information Profile (CHIP)
  - Based on Literature Info
  - Signaled By Chem's In Progress Alert
  - Quality Improved By Voluntary Inputs

Figure 3.

PHASE II - PROBLEM CHARACTERIZATION

- Assess Adequacy of Hazard And Exposure Data

  Approaches:
  - Discussion With Industry
  - Contract Studies
  - Section 8 Rules
  - Section 4 Rules
  - Gov't Testing
- Emphasize Quality of Data

Figure 4.

Section 8(e) reporting, citizen petition Section 21, and selected chemicals from the TSCA Section 5 review program.

The output from Phase I evaluation is called a Chemical Hazard Information Profile (CHIP). This preliminary hazard and exposure data summary relies heavily upon readily available literature. EPA made a wise decision to solicit information from industry by publishing notices of CHIP preparation in the Office of Toxic Substances (OTS) Chemicals in Progress bulletins. By so doing, the Agency has availed itself of the most current and most pertinent data.

Opening the process represents a major step forward in improving the quantity and quality of information used in this important first phase for selection of priority candidates. Without industry input at an early stage, EPA could find itself well down the road in evaluation and control before realizing that its limited technical resources are being wasted on a low priority hazard or risk. Furthermore, industry and the public also are spared the burdens of being misled about potential problems that do not exist, but would be interpreted as real based on incomplete information. Since the CHIP may represent the only official EPA record of an evaluation of a chemical on which no further action is taken, we believe that industry not only should be given the opportunity to provide input, but companies that contribute also should be permitted to review drafts of CHIPs and offer comments before the CHIP is finalized. Otherwise, irreparable harm may come to companies because of erroneous reference documents.

Problem Characterization

Phase II, Problem Characterization, is intended to determine whether the hazard and exposure data are adequate to define the nature and extent of potential problems associated with a chemical (Figure 4). As a follow-up to the preliminary evaluation summarized in the CHIP, an analysis is performed to establish the adequacy of existing data. If costly testing is indicated, collection of exposure data should be considered first to verify that expensive testing is justified by the potential risk.

To the extent that additional data are judged necessary, EPA must select the most cost-effective approach. One of the best approaches to obtain the needed information is through discussions

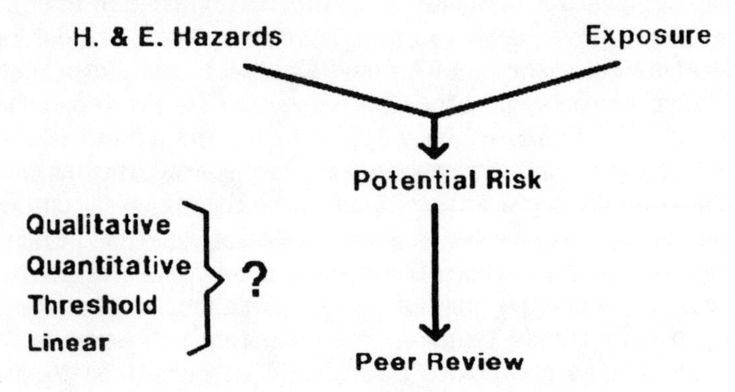
PHASE III - INFO GATHERING AND
RISK ANALYSIS
Assessments:
H. & E. Hazards
Exposure
Potential Risk
Qualitative
Quantitative
Threshold
Linear
?
Peer Review

Figure 5.

with companies identified on the TSCA Inventory as manufacturers of the candidate chemical. This is particularly necessary if exposure information is needed to scope the extent of additional toxicity or environmental effects data required. Other approaches available to the Agency are studies performed by contractors, additional data gathering under TSCA Section 8 rules, TSCA Section 4 testing rules, or testing by a government laboratory.

Regardless of the approach taken to seek added information for characterizing the problem, emphasis must be placed on securing high quality input for risk assessment.

Information Gathering and Risk Analysis

Phase III contains the crucial steps during which information gathering is completed and the all-important risk assessment is made (Figure 5). Health and environmental hazard information is combined with the data on exposure potential to analyze the significance of the risk. The risk assessment may be a qualitative analysis or a quantitative analysis that uses either a linear or threshold model. We are not aware whether EPA uses standardized procedures based on one or a combination of these techniques or whether a case-by-case approach is applied. In any event, EPA's approach to risk assessment should be made public.

Of most importance in Phase III is the need for peer review of the risk assessment performed on each specific substance. The extent of control will be determined by the significance of the risk that has been judged to be associated with the material. Since some risk assessment results are subject to wide variation depending upon assumptions, methodology, and subjective interpretation, draft risk assessments should be made available to the public for critical review. The potential social/economic impact of subsequent control actions and their effectiveness in avoiding unreasonable risks demand thorough scrutiny of the risk assessment. It is in EPA's interest to have the benefit of comments on the draft risk assessment from industry, as well as public interest groups who are capable of participating, before the document is used for setting control rules. Thereby, the possibility for serious conflicts or litigation on ensuing rules can be reduced or eliminated.

PHASE IV – RISK REDUCTION ANALYSIS

- Typical Alternatives:
  - Administrative Controls
  - Hazard Warnings/Labels
  - More Testing
  - Limit Uses
  - Ban
- Balance Risk/Benefits

Figure 6.

PHASE V – RISK MANAGEMENT

- ASSURE NO UNREASONABLE RISK
- Consider Advisory Alternative
- Choose Appropriate Controls
  - Regulating Agency/Law
  - Adequate Time Frame
  - Uniform Enforcement
  - Publicize Audit Procedures

Figure 7.

Risk Reduction Analysis

Having completed the risk asessment, attention is next directed to Phase IV, Risk Reduction Analysis (Figure 6). Here the decisionmaking process begins with the appropriate action to be taken relative to the assessed risk. Various alternatives can be considered. These include administrative controls to change current practices, application of hazard warnings and labels, additional testing to further evaluate potential hazards, and limiting uses or even banning production. Of course, other options are available such as mandatory protective equipment, industrial hygiene and medical monitoring, or substitute products. Whatever control measures are considered, the choice of proper alternatives should be based not only on the assessed risk, but also on the social and economic consequences of the imposed control. In other words, costs and benefits need to be considered to arrive at the proper balance.

Here again, as in previous phases, we encourage EPA to maintain an open dialogue with industry. By so doing, practical choices can be considered based on knowledge and real life experience.

Risk Management

Finally, we arrive at the Risk Management decision in Phase V (Figure 7). In making that decision, EPA must keep in mind it is unreasonable risk against which TSCA is intended to protect. That means making the control decision on the basis of a risk/cost/benefits analysis.

A new control measure being considered by EPA is issuance of a Chemical Advisory to convey information on possible voluntary actions to reduce or prevent risks. Although the concept is appealing in principle, a recommended controls consensus by the affected industry would be required to be workable. We think, however, that the new approach should be given a trial to determine whether, under certain circumstances, it may represent an appropriate alternative.

Another factor concerning the implementation of control regulations which should be considered includes the proper agency for developing the rules and timing. Although EPA's Existing Chemicals Program under the authority of TSCA is equipped to gather information and conduct evaluations, the actual control rules might best be administered under another law or other agency. Regulation may be

PARALLEL INDUSTRY RA/RM ACTIVITIES

EPA	INDUSTRY (MFG./PROCESS)
Section 4	Testing Initiatives
Section 5	TME/PMN/SNUR
Section 6	Controls
Section 8 (a)	Production & Uses
Section 8 (b)	Inventory Chemicals
Section 8 (c)	Allegations
Section 8 (d)	H. & S. Studies
Section 8 (e)	Substantial Risks

Figure 8.

more appropriate by EPA under the Air, Water, or Resource Conservation and Recovery Act (RCRA) laws, or if related to workplace under Occupational Safety and Health Administration (OSHA), or if concerning consumer products under Consumer Product Safety Commission (CPSC), etc. More discussions on this subject are scheduled later in the seminar under the session concerning Government Regulatory Options.

In setting rules for control, the proper time frame also must be considered. A regulation that requires substantial preparation or major changes of widely adopted industry practices must provide ample time for implementation. Flexibility to meet the demands of the rule is essential.

We agree with a strong enforcement program as well in order to assure uniform adherence to the rules. Penalties, however, should be commensurate with the degree of willful violation. Those who come forward with admissions of inadvertent violation should be given due consideration of reduced penalties.

We also would encourage EPA to publicize its auditing procedures so that companies may adopt them for their own internal compliance auditing. Besides allowing companies to comply more completely with the rules, publicizing the auditing procedures strengthens compliance incentives and makes EPA's auditing less complicated.

PARALLEL INDUSTRY RISK ASSESSMENT/RISK MANAGEMENT ACTIVITIES

Having discussed EPA's Existing Chemicals Program, let us turn our attention to industry's applications of risk assessment and risk management.

If we consider the major provisions of TSCA used by the Agency in the risk assessment and risk management process, we find closely parallel activities in industry (Figure 8). In responding to these regulations, companies are conducting their own evaluations of substances that are or may be subject to rules. As a result of these assessments, decisions are made on how best to manage potential risks in internal operations.

In response to Interagency Testing Committee (ITC) recommendations for testing under Section 4, companies have initiated testing on a substantial number of chemicals based on internal risk assessments. Some of these have resulted in voluntary or negotiated testing programs with the Agency. In certain instances, the initiatives have gone even beyond the testing programs prescribed by EPA in order to examine additional potential effects. In other cases, the risk assessments performed by a company or a group of companies have helped demonstrate that no additional testing was needed.

Under Section 5, in order to qualify new chemicals for the same regulatory status as existing chemicals, risk assessments are performed and risk mangement steps are taken prior to submitting petitions for test marketing exceptions (TMEs), in filing premanufacture notifications (PMNs), and in responding to Significant New Use Rules (SNURs). Experience shows that the better the preparation, the less likely EPA will delay, limit, or prevent the manufacture or use of a new substance.

The necessity for companies to perform their own evaluation and institute control measures is obvious for chemicals that are candidates for or are placed under TSCA Section 6 control regulations. Companies must assess the potential risks in their operations and apply appropriate controls to avoid penalties for noncompliance and assure protection for workers and customers.

In relation to all the subsections of Section 8, chemicals nominated for and subjected to regulation are being scrutinized by manufacturers and processors. Exposures during production and uses are being closely examined on Section 8(a) chemicals as well as many Section 8(b) inventory chemicals. Although evaluation is not mandated under the new Section 8(c) allegations recording rule, many companies will conduct an assessment when there are multiple allegations or the allegation of significant adverse effect appears to have substantive support. As the results of health and safety studies on materials covered by Section 8(d) are reviewed, they are analyzed in the context of the known exposure, the adequacy of test data, and sufficiency of precautionary procedures. Last, but by far not least, is the intense risk assessment the law mandates on candidates for Section 8(e) substantial risk notification. When a notice is submitted, a careful risk management review usually is conducted to determine what actions should be taken.

Let there be no underestimation of the level of risk assessment and risk management being triggered in industry by the TSCA regulatory process.

INDUSTRY VOLUNTARY PRODUCT SAFETY

I do not mean to leave the impression that industry's risk assessment or risk management efforts respond only to regulations. Many companies have voluntary product safety stewardship programs that are founded on assessing risks and taking management actions to minimize and eliminate potential identified risks. A few well-recognized examples are worth citing (Figure 9).

Assessments are made on what toxicology tests should be conducted commensurate with workplace and potential customer exposures, known effects, and structure activity relations. When adverse effects are detected in laboratory tests or suspected in humans, similar decisions must be made on the need for epidemiology studies.

For many years, it has been common practice for producers of chemicals to provide Material Safety Data Sheets (MSDSs) with their products. The purpose is to warn their workers and customers of potential hazards and recommend precautions and emergency procedures. The preparation of each bulletin requires a mini-risk assessment and decisions on handling possible risks.

Companies also are engaged in determining acceptable exposure standards, adopting new industrial hygiene practices, and performing medical monitoring. These actions are based on risk evaluations and consequent decisions for appropriate actions.

Based on a combination of many of the above assessments, risk management decisions often will result in changes in manufacturing and/or marketing practices.

SUMMARY AND CONCLUSIONS

As mentioned at the start of this presentation, we are still low on the learning curve with respect to risk assessment and risk management. Neither government agencies nor industry has all the answers. We are each doing the best we can with the knowledge we have and we are each learning by doing. Because we are learning, it is imperative that we establish a dialogue to maximize our learning (Figure 10).

To make an analogy, if we each have a dollar and exchange them, neither of us is any richer. But if we combine the money, we are jointly richer. The same is true with knowledge, only more so.

INDUSTRY VOLUNTARY PRODUCT SAFETY STEWARDSHIP

Examples of RA/RM:

- **Conduct Toxicology/Epidemiology Studies**
- **Issue MSDS**
- **Provide Hazard Warning/Labels**
- **Determine Acceptable Exposure Standards**
- **Adopt IH Practices**
- **Perform Medical Monitoring**
- **Change Manufacturing/Marketing**

Figure 9.

EFFECTIVE RISK MANAGEMENT

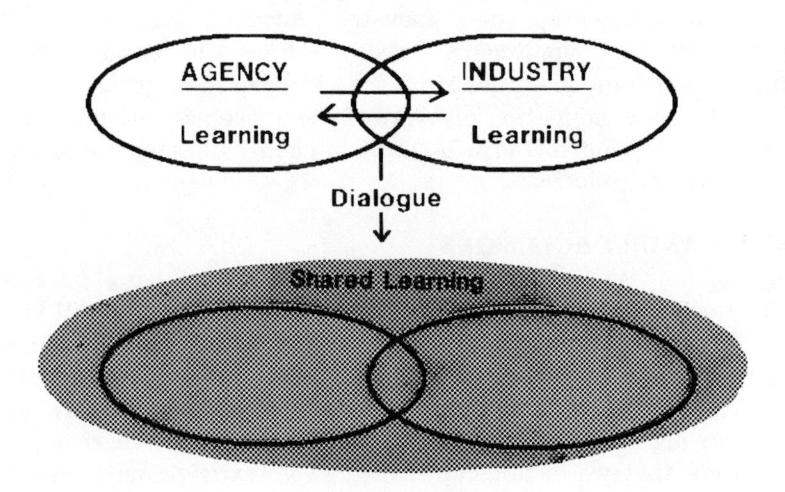

Figure 10.

If we are willing to combine our knowledge through dialogue, we will add the knowledge of others to our own and lose nothing. Furthermore, the synergistic effect of shared learning will provide a multiple increase in the knowledge we both possess.

By increasing our knowledge base, not only does government and industry gain, but so does the public. Risk assessment and risk management are fundamental to all aspects of protecting against unreasonable risk. The success of our product safety efforts depend on fostering dialogue for a partnership in shared learning.

CHAPTER 12

TSCA'S ROLE IN THE OVERALL FEDERAL REGULATORY SCHEME

Robert M. Sussman, Esquire
Covington and Burling

INTRODUCTION

The role of the Toxic Substances Control Act (TSCA) in the overall federal regulatory scheme for protecting human health and the environment is a particularly topical subject.

During the early years of its implementation, TSCA was primarily used to review new chemicals and to collect and evaluate data on existing chemicals. Now, with the creation of the Environmental Protection Agency's (EPA's) Existing Chemicals Program, TSCA is emerging as a major tool for risk management. As TSCA is increasingly used for this purpose, both the Agency and industry have been confronted with difficult questions about the relationship between TSCA and other chemical control laws. How these questions are resolved will have a major impact on federal regulatory policy in years to come.

RISK MANAGEMENT LAWS ENACTED BY CONGRESS

TSCA's testing, data-collection, and new chemical review provisions are virtually without parallel in other federal regulatory laws that apply to industrial chemicals. This is not true, however, of Sections 6 and 7 of TSCA, which contain EPA's authority to ban or restrict commercial activities involving chemicals which present unreasonable risks. Over the years, Congress has enacted a large number of laws which confer similar regulatory powers on EPA and other agencies.

To provide the most obvious examples, foods, drugs, cosmetics, and pesticides have been subject to stringent safety regulation on the federal level for decades. More recently, the Occupational Safety and Health Administration (OSHA) was authorized to control exposure to chemicals in the workplace and the Department of Transportation (DOT) has been directed to assure the safe movement of chemicals in commerce. Also, chemicals used in consumer products are subject to federal safety regulation. Congress passed two laws that apply to household products—the Consumer Product Safety Act and the Federal Hazardous Substances Act—and assigned administration of these laws to the Consumer Product Safety Commission (CPSC). Finally, Congress passed numerous statutes to control the release of chemicals into the environment. These statutes are familiar to all of us. They include the Clean Air Act, the Federal Water Pollution Control Act, the Resource Conservation and Recovery Act, and several others.

INTENT OF CONGRESS IN ENACTING TSCA

Given the scope and variety of other federal statutes, what role should TSCA play in the risk management process? Should TSCA be used as a control mechanism of last resort, to be invoked on those rare occasions when a chemical poses unique risks beyond the reach of other laws? Or should TSCA serve as the main driving force behind the federal effort to regulate chemicals, supplanting more limited laws whenever EPA sees fit?

In addressing these questions, the first source of guidance is the text and legislative history of TSCA itself. When one examines the statute, it becomes clear that Congress addressed the relationship between TSCA and other laws on two levels.

First, Congress exempted certain classes of chemicals from the definition of "chemical substance" and excluded these chemicals from the scope of EPA's authority. The main chemical classes covered by these exemptions are foods, drugs, cosmetics, and pesticides. The basis for exempting these chemicals from regulation under TSCA was the existence of pervasive and stringent safety requirements under other laws.

Comparable exemptions were not established, however, for chemicals that might be controlled by OSHA, CPSC, DOT, or other branches of EPA. Instead of excluding these chemicals from TSCA's coverage, Congress directed EPA to determine the relationship between TSCA and other laws on chemical-by-chemical basis. The procedures and criteria for making these determinations are found in Section 9 of TSCA—a complex and ambiguous provision that has not yet been clearly interpreted.

Section 9(a) addresses EPA's relationships with other federal health and safety agencies. This provision requires EPA to submit a report to another federal agency when it identifies an unreasonable risk which, EPA believes, can be "prevented or reduced to a sufficient extent" under the laws that the other agency implements. EPA must then allow that agency at least 90 days to decide whether to initiate regulatory action to control the chemical. If the other agency proceeds with such action or determines that the chemical does not present an unreasonable risk, EPA cannot commence rulemaking proceedings under TSCA. However, if the other agency declines to take action because of its own resources and priorities, EPA can regulate the chemical under Sections 6 or 7 of TSCA.

Section 9(b) addresses the relationship between TSCA and EPA's other environmental authorities. Where a risk to health or environment can be adequately addressed under another statute administered by EPA, the Agency must proceed under that statute unless it concludes it is in the "public interest" to regulate under TSCA.

As this description illustrates, Section 9 simply does not establish a clear boundary between the coverage of TSCA and other laws. Moreover, Section 9's procedures for coordination between EPA and other agencies involve considerable paperwork and can be quite time-consuming.

One would think Congress could have designed a more efficient and straightforward system for determining where coverage of other laws ends and regulation under TSCA begins. Unfortunately, however, the complexity of Section 9 does not reflect poor legislative drafting as much as the complex (and in ways conflicting) motivations of Congress at the time it passed TSCA.

One the one hand, legislative history demonstrates TSCA was not intended to displace other laws for the protection of health or the environment and that these always would remain the primary mechanism for controlling chemicals to which they applied. On the other hand, Congress was uncertain whether other laws were adequate to address risks that might arise during the manufacture and use of chemicals. It also recognized that other regulatory agencies or EPA offices might be unable to give timely attention to health or environmental concerns identified under TSCA. Given these divergent concerns, Congress crafted a statutory scheme under which other laws would take precedence over TSCA, but only if they contain adequate authority to protect the public and can be invoked in a timely manner. In all other situations, Congress authorized EPA to regulate under TSCA.

EPA'S IMPLEMENTATION OF SECTION 9

Despite the considerable evolution of EPA's Existing Chemicals Program, many questions about the relationship between TSCA and other laws remain unresolved. To date, the Agency has not explained its criteria for invoking TSCA in preference to other health or environmental laws. Moreover, while EPA is actively evaluating several chemicals that could be controlled under other laws, EPA has not yet used the formal report procedure of Section 9(a) to bring its concerns to the attention of other agencies. Instead, EPA has attempted to coordinate informally with other agencies, using behind-the-scenes discussion to determine the respective roles of TSCA and other laws.

For example, EPA recently published an Advance Notice of Proposal Rulemaking (ANPR) on 4,4'-methylene bis(2-chloroaniline (MBOCA), a chemical for which OSHA had attempted to develop an occupational safety and health standard during the 1970s. In its ANPR, EPA indicated it had informally consulted with OSHA and the two agencies had concluded that TSCA represented the best vehicle for future regulatory action. The ANPR did not explain why regulation of MBOCA, which is of concern because of occupational risks, can best be accomplished by EPA rather than OSHA.

EPA's recent activities involving 4'4-methylenedianiline (MDA) provide another example of informal coordination between EPA and OSHA. After EPA designated MDA for priority action under Section 4(f) of TSCA, the two agencies simultaneously published ANPRs requesting comments on the need for regulatory action. To date, EPA has taken the lead role in evaluating MDA, and it presently appears that rulemaking proceedings will be commenced by EPA rather than OSHA. It will be of interest to learn why EPA believes that MDA, which principally raises workplace concerns, is an appropriate candidate for regulation under TSCA rather than the Occupational Health and Safety Act. It will also be of interest to compare any workplace requirements for MDA that are proposed by EPA with occupational safety and health standards issued for comparable chemicals by OSHA.

Asbestos provides a third example of EPA's efforts to coordinate with other agencies. OSHA issued a temporary emergency standard reducing exposure to asbestos in the workplace. Simultaneously, EPA is developing a proposed rule under TSCA which would severely restrict the use of asbestos in certain major applications. EPA's proposal is apparently motivated by occupational health concerns. EPA has said it is informally coordinating with OSHA and other agencies. Nevertheless, a real prospect exists that, in the near future, two different agencies will be simultaneously conducting proceedings to reduce the occupational risks posed by a single chemical.

CONCERNS RAISED BY EPA'S CURRENT APPROACH TO COORDINATION WITH OTHER AGENCIES

To some degree, it is inevitable that chemicals which raise workplace or consumer product concerns would receive scrutiny by EPA under TSCA. As a result of its activities under the testing and data-gathering provisions of the Act, EPA will often devote attention to chemicals whose primary exposure potential is in the workplace or in consumer products. If EPA concludes these chemicals warrant control, consideration of action under TSCA would seem natural, particularly if OSHA and CPSC are busy with regulatory activities of their own.

It is also understandable that EPA would shy away from the referral procedure of Section 9 and seek to coordinate with other agencies on a more informal basis. EPA may feel that preparing a Section 9(a) report is too time-consuming and that is is more efficient to consult informally with other agencies. As its use of joint ANPRs suggests, EPA may also wish to preserve its flexibility and select a control strategy as late in the regulatory process as possible.

Nevertheless, while informal coordination may seem attractive from an administrative standpoint, it has disturbing long-term implications which should concern both industry and regulatory agencies. As EPA expands its activities under TSCA, it will begin to regulate chemicals that, up to now, have been the exclusive province of other agencies like OSHA and CPSC. With time, areas like occupational safety and health may fall within the jurisdiction of two agencies, each implementing different statutory schemes and bringing to bear different types of expertise.

One consequence of this development could be conflicting regulatory approaches to essentially identical health or environmental concerns. There are important differences between the substantive requirements imposed by TSCA and those which apply under other laws. A Section 6 rule, for example, requires a finding that a chemical may present an "unreasonable risk of injury"—a concept that requires EPA to balance a chemical's potential for adverse effects against its social and economic benefits. In promulgating a workplace standard, by contrast, OSHA must simply find that a chemical presents a "significant risk" at prevailing exposure levels. No balancing of costs and benefits is required or even permitted.

Moreover, the remedies available to EPA under TSCA are more varied and far-reaching than the control measures authorized under other statutes. For example, an occupational standard issued by OSHA must reduce workplace exposure to the lowest level technically and economically achievable. Section 6, by contrast, permits EPA to ban a chemical, restrict its use in some or all applications, impose a ceiling on total production, or simply require cautionary labeling.

Because of the differences betwen TSCA and other laws, chemicals that raise similar health or environmental concerns could be treated differently depending on whether they are subject to EPA rules under Section 6 or requirements imposed by other agencies. For example, OSHA would regulate a chemical posing a risk of chronic health effects by developing a standard which establishes a uniform Threshold Limit Value (TLV) for all workplaces where the chemical is manufactured or used. EPA, on the other hand, might promulgate a Section 6 rule which eliminates some or all applications of the same chemical on the ground that they present an "unreasonable risk." It is hard to justify such differences in regulatory treatment where they do not correspond to real differences in risk.

The absence of clear boundaries between TSCA and other laws frustrates rational planning by industry. Where there is a high likelihood that a particular agency will regulate a chemical and that agency's policies and procedures are well-understood, companies may anticipate regulatory requirements by modifying control techniques or altering marketing strategies. Anticipating regulatory needs in this manner is impossible, however, if companies cannot predict which agency will take the lead role in controlling a chemical and what approach that agency will follow.

Finally, uncertainty about the relationship between different laws undermines the government's own ability to function effectively. The staffs of different agencies may devote significant attention to a specific chemical before they realize they are duplicating each others' activities. Moreover, one agency may devote extensive resources to issues that would be irrelevant if another agency initiated regulatory action. For example, EPA routinely asks its contractors to evaluate the availability of substitutes for chemicals under consideration for Section 6 action. The availability of substitutes is clearly a legitimate issue under the balancing test contained in Section 6. However, this issue will become unimportant if the chemical is ultimately regulated by OSHA, which does not employ a cost-benefit analysis.

SUMMARY AND CONCLUSIONS

I have no easy solutions to recommend for the problems I have described. Section 9 of TSCA is admittedly complex and does not provide clearcut guidance on the relationship between TSCA and other laws. Moreover, given the comprehensive nature of the Act, some overlap between TSCA and other laws is probably unavoidable.

Nevertheless, there are dangers in continuing to determine the relationship between TSCA and other laws on an ad hoc basis, without any effort to examine long-term implications of decisions on particular chemicals. Accordingly, while EPA's interest in preserving its flexibility is understandable, the public may benefit by a clear set of guidelines defining the relationship between TSCA and other laws. Section 9 may not provide all the answers, but it does embody two important principles: first, other laws should take precedence over TSCA unless they are inadequate to address a particular risk and cannot be invoked in a timely manner; and second, the choice among different laws should reflect rational policy considerations and should be made in a public, open manner. By adhering to these basic principles, EPA can improve the operation of its Existing Chemicals Program.

CHAPTER 13

USE OF QUANTITATIVE RISK ASSESSMENT IN REGULATORY DECISIONMAKING UNDER FEDERAL HEALTH AND SAFETY STATUTES

Peter Barton Hutt, Esquire
Covington and Burling

INTRODUCTION

Risk assessment and risk management are relatively new terms in the field of government health and safety regulation. These concepts, however, have their origins in antiquity. Indeed, government regulation to protect public health and safety represents perhaps the oldest form of government control of commercial enterprise.

HISTORY OF FOOD AND DRUG REGULATION

The Ancient World

The earliest recorded regulation of health and safety was designed to protect the public against unsafe food. These controls, in the form of the dietary laws handed down by Moses, were a direct reflection of judgment based upon human experience. They prohibited, for example, the use of pork or the meat of any diseased animal, undoubtedly as a result of the observation that consumption of these products frequently led to human disease.

Roman civil law included a broad edict against any form of commercial fraud. Adulteration of food or drugs constituted a civil offense, subject to government prosecution, and resulted in such punishment as condemnation to the mines or temporary exile.

It is not surprising that early laws were designed to protect against food and drug adulteration. Pliny the Elder, writing in the 1st century A.D., documented numerous examples of widespread adulteration of the food and drug supply. Pliny deplored the "avarice and luxury" of the merchants, who "spoil everything with fraudulent adulterations":

> So many poisons are employed to force wine to suit our taste—and we are surprised that it is not wholesome!

Other authors in ancient Greece and Rome reported similar problems.

England

Early English statutes were directed primarily at assuring that a given quantity of food would be sold at a given price. Very quickly, however, it became apparent that the price of food could not be controlled without adequate regulation of food quality. Accordingly, as early as 1266, Parliament enacted the Assize of Bread, which prohibited the sale of any staple food product "that is not wholesome for Man's Body." That statute remained the law of England until 1844. It is apparent, moreover, that more modern statutes subsequently enacted in England and later in the United States have not improved in terms of the legislative directive to prohibit unsafe food.

Throughout the history of the Assize of Bread, however, enforcement remained subject to human experience and judgment—the same inexact factors on which Moses relied in handing down his dietary laws. Risk assessment at that time was based solely on trial and error.

The Dawn of Toxicology

Hippocrates is credited as the founder of medicine because he related all health and disease to natural, rather than supernatural, causes. But it was Paracelsus, an enigmatic alchemist writing in the first half of the 16th century, who first articulated the concept of a dose-response curve:

> All substances are poison; there is none which is not a poison. The right dose differentiates a poison and a remedy.

For all its insight, however, this did not advance risk assessment or regulatory decisionmaking. Paracelsus correctly pointed out there is a line which divides a safe and an unsafe dose, but he offered no criteria for determining how or where to draw that line. Thus, he merely restated the problem that existed from the time of Moses. It took almost another four centuries before the dose-response curve assumed its present importance.

The Dawn of Chemistry

Although even Pliny and his contemporaries could find gross adulteration of food and drugs, it was another 15 centuries before analytical chemistry techniques were developed to determine the presence of adulterants. Led by the landmark work of Robert Boyle in the second half of the 17th century, new tests were developed for food and drug adulteration. As the field of chemistry matured, detection of adulteration advanced beyond individual opinion—based on uncertain tests such as burning, sight, taste, and smell—and relied instead upon objective and reproducible criteria.

These new chemical tests, in turn, led to increased public concern about the safety of the food supply. A 1760 pamphlet warned about the "destructive ingredients . . . to health" used in "eatables and drinkables." This concern culminated in the publication, in 1820, of Frederick Accum's Treatise on Adulterations of Food and Culinary Poisons. In that treatise, Accum described, in detail and at length, the numerous kinds of adulteration practiced on both food and drugs, and the various methods available to detect them. He documented widespread problems throughout the food and drug supply, many of which were "highly deleterious" to the public health. His treatise was an immediate and worldwide success, and it galvanized scientists into conducting further studies and legislators into the enactment of modern laws to protect the public health and safety.

The Rise of Public Health Laws

Chadwick, working in England, documented the importance of sanitation in protecting the public health. Following that lead, Shattuck issued his monumental Report of the Sanitary Commission of Massachusetts in 1850, documenting the decrease in average life expectancy at birth in America's large urban centers because of insanitation. Shattuck recommended enactment of statutes to control the food and drug supply as part of a broad public health approach. Following that report, boards of health were established throughout the country and regulatory controls were strengthened at the state and local level.

Early American Regulatory Statutes

The American colonies imported the same laws that existed in England to protect the colonists' public health and safety. After the American Revolution, these laws were reenacted and made permanent. Throughout the 19th century, protection of public health and safety was considered largely a state and local matter, and thus each jurisdiction had its own specific enactments.

As early as 1813, however, Congress enacted the first federal regulatory statute to protect the public against unsafe and ineffective smallpox vaccine. That statute lasted only until 1822, when it was repealed in an early wave of regulatory reform. In 1848, Congress enacted a new law to protect against the importation of unsafe or ineffective drug products, a law that lasted until it was superseded in 1926. And in the late 1800s Congress passed a series of laws designed to protect against the spread of illness through diseased livestock. In 1902, following the death of 11 children in St. Louis from a contaminated vaccine, Congress enacted yet another statute to require the premarketing testing and approval of all biological products, a statute that remains in force to this day.

1906 Food and Drugs Act

From 1879 to 1906, Congress investigated the safety of the food and drug supply and debated at length whether broad federal controls should be enacted to regulate it. In order to dramatize problems with the food supply, Dr. Harvey W. Wiley, then Director of the USDA Bureau of Chemistry, which later became the Food and Drug Administration, established a "poison squad" of 12 male employees of the Bureau. During 1902-1904, feeding experiments were conducted on these 12 young men using the following five preservatives then found in the American food supply: boric acid and borax, salicylic acid and salicylates, sulfurous acid and sulfites, benzoic acid and benzoates, and formaldehyde. Detailed records were kept on all subjects. The results, published in five parts during 1904-1908, provoked widespread interest throughout the country and unquestionably contributed greatly to the enactment of the 1906 law.

It is important to understand the significance of the 1902-1904 human feeding experiments on these five classes of food preservatives. As far back as the ancient world of Greece and Rome, animal experimentation was commonly practiced. Indeed, numerous erroneous theories about human anatomy were perpetrated for centuries as a result of reliance upon animal experimentation unconfirmed by observation of human experience. Through the end of the 1800s, however, animal experimentation remained essentially an ad hoc matter. Physicians and scientists, aware of the differences in anatomy between animals and humans, distrusted animal testing as a model for judgment about human safety. There were, moreover, no uniform animal colonies on which to conduct experiments, and the concept of untreated controls and statistical evaluation had not yet been advanced. Thus, although individual scientists stressed the use of animal testing for safety evaluation in the late 1800s, most risk assessments relied on human feeding studies.

Because of controversies about whether the use of the preservatives tested by Wiley and other food ingredients violated the safety provisions of the 1906 Act, President Theodore Roosevelt appointed a five-man Referee Board of Consulting Scientific Experts in February 1908 to rule on these issues. Chaired by President Ira Remsen of Johns Hopkins University, this was the first scientific regulatory advisory committee in history.

Faced with the single human feeding study on sodium benzoate, the Referee Board promptly determined that additional scientific studies were needed. Rather than rely on animal experiments, however, three of the members of the Referee Board returned to their academic institutions to conduct three additional human feeding studies on students at Columbia, Northwestern, and Yale. Following these additional studies, they reconvened and determined that the existing levels of sodium benzoate used in the food supply were safe.

Development of Animal Toxicity Testing

From Moses to Wiley, as we have seen, safety determinations were based on direct human observation. Although some testing was conducted on animals, it was regarded as unreliable and insufficient for final safety determination.

At the same time Wiley was conducting his human feeding experiments, however, the foundation was being laid for modern animal toxicity testing. During the 1910s, American scientists began to develop colonies of pure inbred strains of rodents, on which an enormous body of toxicological knowledge was obtained. As Wistar, Sprague-Dawley, Osborne-Mendel, and other rat strains became available, large-scale animal toxicity testing for the first time became feasible. Although the history of the development of these colonies has not yet been written, it is apparent the reproducibility of results on these animal colonies, and thus the ability of physicians and scientists to compare results with human experience, ultimately overcame earlier scientific distrust of the use of animals as models for human toxicity. Within a short period, animal testing became an accepted part of toxicology. Throughout the 1930s, animal toxicity testing was common.

Use of Safety Factors

As long as safety evaluation remained primarily based on direct observation of human experience, safety determinations were almost completely judgmental. There could be no attempt at quantification of risk. One searches in vain, for example, for analytical evaluation or even discussion of the elements that were included in the determinations by Wiley and the Referee Board about the safety of sodium benzoate during 1904-1909.

With the advent of controlled animal experimentation, however, operational definitions of safety became feasible for the first time. The insight of Paracelsus that toxic response is a function of dose could at long last serve as the basis for regulatory decisions.

During the 1940s, Food and Drug Administration (FDA) toxicologists and pharmacologists, led by Dr. Arnold Lehman, developed an informal safety factor for regulatory purposes. Based on the work of colleagues in England and America, FDA concluded there was potentially a ten-fold greater sensitivity of humans than test animals, and potentially a ten-fold difference in susceptibility among members of the human population, providing a safety factor of 100. A safe level for humans (now called the "acceptable daily intake" or ADI) was determined by dividing the lowest "no effect level" (now called the "no-observed-effect-level" or NOEL) by this safety factor of 100.

The 100-1 safety factor was undoubtedly more intuitive than scientific when it was first suggested. Indeed, FDA scientists emphasized during the 1950s, even after they had been using it for some time, that human safety determinations could not be made solely on the basis of a safety factor, but must include scientific judgment as well. Since then, some have sought to demonstrate that the 100-1 safety factor has a relatively strong basis in the biological sciences, and this factor remains in widespread use for all toxic effects shown to have a threshold.

Special Problem of Carcinogens

Animal toxicity testing for carcinogenicity occurred throughout the 1930s, and as early as 1945 Berenblum pointed out that "it is frequently necessary to evaluate carcinogenic potency and neoplastic response on a quantitative basis." Even at that early date, however, knowledge about the uncertainty of a threshold for carcinogens caused FDA to impose more stringent safety factors. The Agency initially adopted a 2000-1 and then a 5000-1 safety factor for carcinogens. By 1950, however, FDA concluded no safety factor could be justified for a carcinogen, and imposed a zero tolerance for any carcinogen in the food supply. In that year, FDA banned two nonnutritive sweeteners (dulcin and P-4000) from the food supply on the basis of animal feeding studies demonstrating carcinogenicity. When this policy of zero tolerance for carcinogens was codified into statutory law as part of the Food Additive Amendments of 1958 in the famous Delaney Clause, FDA acquiesced on the ground that a specific anticancer provision merely reflected existing Agency policy and thus made no change in the law.

As a practical matter, a zero tolerance for carcinogens in the food supply meant that no carcinogen could purposely be added and that no contaminant could be present at the level of sensitivity of the detection methodology available at that time. At the beginning of the 1950s, detection methodology was generally sensitive in the range of 20-50 ppm. By the end of the 1960s, however, detection methodology was sensitive in the very low ppb range.

Regulation of Carcinogenic Animal Drugs

Prior to 1958, FDA had approved the use of diethylstilbestrol (DES) in livestock and poultry, under conditions which the Agency believed would not result in residues in human food. Following enactment of the Delaney Clause, the Agency refused to grant new approvals of DES, although it was precluded from revoking old approvals. Congress, therefore, enacted the "DES proviso" in 1962 explicitly requiring FDA to approve DES for use in food-producing animals as long as residues could not be found in the resulting human food under detection methodology specified by FDA. FDA adopted an official analytical method sensitive to two ppb for purposes of enforcing this provision.

In 1972, however, radioactive tracer studies found residues of DES in livestock even after substantial periods of withdrawal of the drug. FDA promptly withdrew approval of DES. It was immediately apparent to the Agency, however, that the ramifications of the DES crisis went far beyond the specific substance involved. If DES could be found in livestock, at low levels, after a lengthy period of withdrawal, it was apparent any other carcinogenic animal drug or indeed any other carcinogenic feed ingredient or contaminant could also be found. Thus, over a weekend, the entire basis on which FDA had been regulating carcinogenic substances in the diet of animals was destroyed. FDA no longer had any rational basis for setting the level of sensitivity for an acceptable detection methodology for carcinogenic substances in livestock, because it was now certain those substances would be present at some finite level even if they could not be detected. Since the Agency believed no level of a carcinogen could be regarded as safe, its only option was to ban all carcinogenic substances from animal feed and animal drugs—an impractical, if not impossible, result.

FDA 1972 Adoption of Low-Dose Extrapolation for Quantitative Risk Assessment

The concept of quantifying human risk on the basis of animal feeding studies had already been explored by scientists for 25 years. During this time, however, it had remained an entirely academic inquiry, unrelated to regulatory decisionmaking.

In 1945, for example, Berenblum published a system for grading carcinogenic potency of chemicals. Iversen and Arley developed a model beginning in 1950 based on a one-hit theory of carcinogenicity. Fisher and Holloman advanced a multicellular concept in 1951. Nordling and Stocks proposed multistage theories in 1953, which were modified by Armitage and Doll in 1954.

In 1961, Mantel and Bryan not only advanced their own log-profit mathematical model for determining carcinogenic risk through low-dose extrapolation, but urged that a risk of 10^{-8} (one in 100 million) be regarded as a "virtually safe dose" for purposes of human exposure. In 1970, indeed, Gross and Mantel published a paper applying the Mantel-Bryan model to the regulation of methyl salicylate, which involved reproductive rather than carcinogenic effects.

Thus, by 1972 there was a substantial body of theoretical literature on low-dose extrapolation from animal toxicity studies for purposes of determining human risk from carcinogens. Within days after FDA withdrew approval of DES in 1972, Dr. Richard Lehman of the FDA Bureau of Veterinary Medicine walked into the office of the FDA Chief Counsel, described the Mantel-Bryan method, and suggested it might be useful in handling problems like DES in the future. The applicability of this concept to regulating carcinogenic animal drugs was immediately apparent to the Chief Counsel, particularly because no other method of regulation remained feasible. The task of preparing an appropriate regulation incorporating the concept was begun promptly, but the complexity of the regulatory and scientific issues and the need for broad consensus within the Agency resulted in a delay of almost a year before a proposed regulation was published in the Federal Register. The July 1973 proposed regulation represented the first regulatory use of low-dose extrapolation for purposes of quantifying human risk from a carcinogen and establishing a "safe" level of human exposure.

FDA proposed to require use of the mathematical model advanced by Mantel and Bryan in 1961 to determine the level at which a carcinogenic residue in human food would present no more than a one in 100 million lifetime cancer risk to the exposed individual. The manufacturer of any carcinogenic animal drug was required to develop an analytical method sufficiently sensitive to determine a residue of the drug in food at the level which represented that risk.

This approach is commonly referred to as the "sensitivity of the method" or "SOM" concept. If a residue could not be detected in food at that level, use of the carcinogenic animal drug would be approved. Thus, through this approach, FDA directly defined the concept of a "significant" carcinogenic risk using quantitative risk assessment based upon low-dose extrapolation.

After extensive public comment, FDA revised the proposed regulation and promulgated it in final form in February 1977. The final regulation adopted one in one million as an insignificant risk level and substituted the "improved" Mantel-Bryan mathematical model published by Mantel in 1975. That final order was overturned in the courts on the procedural ground that no opportunity had been given to the public for comment on the 1975 Mantel-Bryan mathematical model. The entire regulation was reproposed in March 1979 with still further changes. A final regulation is presently in the last stages of approval within the Agency. Since 1972, however, the regulation of carcinogenic animal drugs by FDA has proceeded using the basic concept of low-dose extrapolation to determine an insignificant level of human risk.

Proliferation and Refinement of Mathematical Models

If it achieved nothing else, the July 1973 proposal published by FDA galvanized biological scientists into considering still further refinement of mathematical models for low-dose extrapolation to reflect the most recent knowledge and hypotheses about the carcinogenesis process. A vast body of literature has developed on this subject. Indeed, it has become a scientific subspecialty of its own. Because of the great improvement in the methodology and the realization that further insights are likely to come rapidly in the future, FDA has concluded it should not specify any mathematical model or quantitative risk assessment methodology by regulation, but must instead leave such details for constantly changing administrative guidelines.

Use of Low-Dose Extrapolation by EPA Beginning in 1976

The Environmental Protection Agency (EPA) was created by an Executive Order issued in 1970 to consolidate regulation of environmental safety, including pesticides. During the first half of the 1970s, EPA proposed to take action against three important pesticides found to be carcinogenic in test animals: DDT, aldrin/dieldrin, and chlordane/heptachlor. For several years, administrative proceedings presented the same difficult questions of regulating carcinogens that FDA had already faced in the area of animal drugs. In an attempt to codify its developing policy for carcinogens, EPA published general cancer principles in May 1976, which included the use of low-dose extrapolation to assess the level of human risk through quantitative risk assessment. Unlike the FDA proposal of 1972, the EPA principles of 1976 did not adopt either a specific mathematical method or a specific level of acceptable human risk. Beginning in 1976, however, the EPA Carcinogen Assessment Group (CAG) has undertaken quantitative risk assessments on a wide variety of chemicals, including pesticides and air and water pollutants. As one example, EPA used this approach in 1980 to establish an acceptable level of one ppm of N-nitroso contaminants in pesticide products, which it calculated to represent less than one in one million individual lifetime risk of human cancer.

Other FDA Uses of Quantitative Risk Assessment in Regulatory Decisions

In addition to the use of quantitative risk assessment to regulate carcinogenic animal drugs beginning in 1972, FDA gradually realized that the large number of carcinogens throughout the food, drug, and cosmetic supply required application of quantitative risk assessment whenever the regulation of a carcinogenic substance was involved. This realization has not come easily or quickly, but has progressed gradually over the past decade.

Based on findings of carcinogenicity, FDA had banned acrylonitrile for use in beverage containers in 1977. Following a court decision in 1979 allowing FDA to ignore insignificant levels of risk, however, the Agency has begun to determine acceptable risk levels for acrylonitrile, polyvinyl chloride, 2-nitropropane, and other food packaging additives which pose extremely small levels of human risk as a result of their intended uses.

Based on a 1978 finding of carcinogenicity, FDA proposed requirement of a cancer warning on the labels of all hair dye products containing the substance 2,4-DAA (4-MMPD). In the face of a quantitative risk assessment showing an individual lifetime cancer risk of less than one in one million, FDA nonetheless persisted and promulgated a final regulation requiring a more stringent warning than Congress has required for cigarettes. Upon confronting a challenge filed by the cosmetic industry in the courts, however, FDA signed a consent order staying the regulation, remanding it for reconsideration, and stipulating that any further rulemaking would utilize "scientifically accepted procedures of risk assessment" to determine whether the substance presents "a generally recognized level of insignificant risk to human health." Since then, FDA has not proceeded further on this matter. In 1980, FDA approved the use of a carcinogenic hair dye, lead acetate, which had been shown to present less than one in one million individual lifetime cancer risk. When consumer advocates objected to this action, the Agency reiterated its decision in 1981. No court action was brought to challenge the final decision.

The regulation of color additives has been plagued, throughout this time, by the detection of carcinogenic contaminants in many of the pure dyes. FDA disapproved a number of color additives on this ground, and in 1978 proposed to disapprove one that it had previously approved when it discovered beta-naphthalamine contamination. Industry objected to this action on the ground that the risk presented by this contaminant was less than one in one million individual lifetime cancer risk. No further action has been taken by FDA on this proposal.

In 1982, indeed, the Agency published an advance notice of proposed rulemaking setting out a new "constituents policy" to handle these matters. Under this approach, any nonfunctional constituent in products regulated by the Agency would be regarded as acceptable if it posed only an insignificant risk of cancer, utilizing scientifically acceptable procedures for quantitative risk assessment. Based on that policy, FDA has subsequently approved a number of color additives and food additives containing nonfunctional carcinogenic contaminants. The policy itself was challenged in the courts and has been upheld in a decision in the United States Court of Appeals for the Eighth Circuit. The court ruled that a cancer risk of between one in 30 million and one in 300 million is "safe." In the interim, EPA has used the FDA constituents policy to approve a pesticide.

Use of Quantitative Risk Assessment by CPSC and OSHA

For policy and philosophical reasons unrelated to their statutory authority, the Consumer Product Safety Commission (CPSC) and the Occupational Safety and Health Administration (OSHA) were initially reluctant to incorporate any form of quantitative risk assessment in their regulatory decisions. In the Aqua Slide "N" Dive case in 1978, however, the court rejected a warning required by CPSC on swimming pool slides, on the ground that the risk involved was less than the risk an average person has of being killed by lightning. In the Benzene decision in 1980, the Supreme Court interpreted the Occupational Safety and Health Act to apply only to significant risks to human health. Since then, both agencies have begun to incorporate quantitative risk assessment into their regulatory decisions. In its 1983 proposal to regulate ethylene oxide, for example, OSHA relied heavily upon a quantitative risk assessment.

STRENGTHS AND WEAKNESSES OF QUANTITATIVE RISK ASSESSMENT

It is clear from developments during the past 12 years that quantitative risk assessment represents the principal focus of regulatory decisionmaking for carcinogens for the foreseeable future. Like all regulatory tools, however, it must be recognized that this approach has both strengths and weaknesses.

The major strength of quantitative risk assessment is that it permits regulatory agencies to base decisions on an evaluation of potential human risk involved. Prior approaches, such as safety factors or the level of detection of the available analytical methodology, relied instead on the artificial concept that substances are either safe or unsafe.

The need for evaluation and comparison of the level of risk presented by carcinogens has become more important as regulatory agencies recognize it is impossible to eliminate all carcinogens from the environment. FDA began to focus on this problem after the DES crisis of 1972, and realized its full implications when EPA revealed, in the mid-1970s, that the nation's water supply contains more than ten known carcinogens. Virtually all food is processed with or contains water drawn from that water supply. By 1979, FDA recognized most food contains one or more carcinogenic constituents. Today, it is acknowledged that no human could survive solely on a diet entirely free of carcinogens.

It is increasingly apparent, indeed, that everything we consume modifies the carcinogenic process in one way or another. It is likely, in fact, that particular substances may act differently under varying circumstances, either to promote or to retard cancer. Under these complex circumstances, which are not yet understood, a simplistic distinction between safe and unsafe substances is impossible.

Quantitative risk assessments permit regulators to differentiate between significant and insignificant risk and to tailor regulatory decisions to the goal that has existed throughout the centuries—protection of public health and safety. Thus, its appeal to regulatory agencies is overwhelming.

At the same time, the obstacles to quantitative risk assessment remain formidable. For example, use of this technique depends on the availability of adequate and reliable animal data and human exposure information. Uncertainty about the process of carcinogenesis in humans has led to the development of mathematical models that are extraordinarily conservative. Conservative assumption is piled upon conservative assumption, with the result that the "upper bound risk" that is usually calculated vastly overstates realistic risk potential. Those calculated risks, moveover, are easily misunderstood as "hard numbers" by the general public. Quantitative risk assessment is also frequently misapplied to secondary carcinogens as well as to primary carcinogens, largely because of the extreme difficulty at this time in distinguishing between the two.

Quantitative risk assessment in itself, moreover, does not predetermine the outcome of regulatory decision. As the National Academy of Sciences recently emphasized, risk management (i.e., a regulatory decision) is entirely separate and distinct from risk assessment (i.e., an evaluation of the level of risk presented by a particular substance). Numerous policy and practical judgments are required to convert a risk assessment into a regulatory decision.

In spite of these weaknesses and drawbacks, however, quantitative risk assessment remains the best way currently available to organize and analyze toxicity data for purposes of making difficult regulatory decisions. Although this is not yet a mature science, enormous progress has been made since FDA initiated the process in 1972, and it can be anticipated that substantially greater progress will be forthcoming in the future as more scientific information becomes available about the process of carcinogenesis itself.

SUMMARY AND CONCLUSIONS

The future of all health and safety regulation is inseparable from the future of quantitative risk assessment. Since we now understand it is impossible to have a risk-free environment, the only realistic approach for government regulatory agencies necessarily includes an assessment of existing risks and the reduction in those risks made feasible through alternative regulatory actions.

Perhaps the most difficult task facing government and industry today is to reeducate the American public to understand and expect that no element of our environment, including the food supply, can ever be entirely safe. Everything we consume or touch represents some degree of risk. Far from promising regulatory action that guarantees safety, regulatory agencies can promise only to reduce risks to the lowest level feasible or, in any event, to an acceptably small level.

It will take years before this new understanding permeates our society. In the interim, severe misunderstandings and dislocations have occurred and will continue to occur. Unrealistic expectations about public protection cannot easily or quickly subside. This educational process, on which the success of quantitative risk assessment and future regulatory approaches depends, is the single most important objective for the future.

CONCLUSIONS

SUMMARY

Carl W. Umland
Exxon Chemical Americas

The papers presented at the Chemical Manufacturers Association Risk Management Seminar provided a great deal of thought-provoking comments which suggest we all have our work cut out for us. To sum up comments heard from the first panel of speakers I will offer some of the most significant points.

Marcia Williams, Deputy Assistant Administrator, Office of Pesticides and Toxic Substances of the Environmental Protection Agency (EPA), did an excellent job of describing the existing chemicals program at EPA with sufficient examples I think to help understand that program. I heard expressions placing emphasis on steps to increase early public awareness of EPA thinking and activity. I also heard an overt philosophical statement cum policy that the Toxic Substances Control Act (TSCA) is a broad statute which allows entry into any regulatory area needed including the workplace and consumer products (with close coordination among agencies, of course). We were also given a message that greater focus and a better job of risk management of "problem chemicals" would be provided by attacking those problems across a broad base. And, last but not least, we heard a plea for help in finding ways to publicize the positive accomplishments of both EPA and industry in the effective implementation of TSCA to date.

J. Clarence Davies dealt effectively with the semantics of discussing risk and its control. He called for policy interactions with scientific assessment, but I was impressed that there is room for accommodation of views on this issue because we have common objectives. His point regarding public expectations was telling—the public does not want to exchange one unreasonable risk for another, it does want freedom from fear. The public is presently confused and often misinformed. Solutions suggested were controls of exposures across media.

Deems Buell portrayed a strong network of activity in industry which offers know-how and tools for improving the current situation.

Overall, we heard a rational and reasonably balanced expression of a new willingness to move forward in common cause. To be sure, there will be differences on precisely how to get to Valhalla along the way, but I sensed that better understanding may be moving all industry beyond the name-calling and posturing stage to the point where we can anticipate rational and effective progress and improvement in perceptions.

CONCLUSIONS

WRAP UP COMMENTS

J. Ronald Condray
Monsanto Company

The panel moderators for the "Risk Management of Existing Chemicals" Seminar summarized the discussions of the various panels. I will not expand on their discussions, but rather, will reflect a little on the comments of our guest speakers.

Dr. Jack Moore, Assistant Administrator of the Environmental Protection Agency's (EPA) Office of Pesticides and Toxic Substances, and Senator Durenberger, the chairman of the Senate Subcommittee on Toxic Substances and Environmental Oversight, both focused on several issues that the chemical industry, government regulators, and public interest groups must keep in mind.

1. We must accept the fact that public perception regarding toxic substances is a reality that must be dealt with. Congress, government regulators, and industry must be prepared to respond in a positive fashion to concerns the public has regarding control and management of the risks associated with toxic substances. While these may be perceptions, they are nonetheless motivators to action, and a stonewall approach on the part of either Congress, the regulators, or industry will do nothing but fuel the fire of those concerned, that something is being hidden in the area of toxic substances. Mr. Bill Simeral, past chairman of the Chemical Manufacturers Association's (CMA) Board of Directors, has stated publicly in several forums that fear of chemicals is one of the most important elements that we in industry must address. I am convinced the more we educate and communicate with the public, the better off we will be in overcoming what, in the eyes of industry, is a perception rather than a reality.

2. We must accept the fact that the Toxic Substances Control Act (TSCA) is a very broad act and deals with a number of sectors beyond the classical chemical control parameters. Such areas as groundwater contamination, workplace exposure, and fish kills in the Great Lakes, all deal with toxic substances issues. We must strive to find effective ways to bridge responsible management of these toxic issues within the regulatory community. TSCA can effectively deal with toxic issues in areas previously considered the exclusive realm of other environmental laws and regulations. There must be innovative ways of using the authorities of TSCA coupled with the authorities of other environmental control laws to harmoniously accomplish the goals of protecting human health and the environment.
3. Both guest speakers challenge the industry to reassess the handling of confidential business information (CBI). Industry must give the public assurances that bad data is not being hidden behind CBI claims. We need to identify information that is truly proprietary and confidential and be more willing to publicly share other information that is clearly not confidential or of significant proprietary value.
4. The fourth area our guests addressed was the area of risk assessment versus risk management. Industry and government was challenged to do less talking about risk assessment and more action regarding risk management. It's true, risk management under TSCA is a difficult task and one that both EPA and industry are only recently learning to understand and control. Nonetheless, it's clear from our guest speakers' perspective we must move on to risk management on a priority basis.
5. Lastly, Senator Durenberger raised this point: we must be aware our actions have a global context. There is a need to harmonize risk assessment and risk management schemes around the world in order to reduce the potential trade barriers.

The challenges offered by our guest speakers cannot be ignored. Congressman Florio has introduced legislation to address some of these concerns and Senator Durenberger is seriously considering TSCA amendments. We must continue to discuss these issues and work together with all parties to overcome what I believe is the basic overriding issue of all, the public fear of chemicals. I believe this seminar made a significant and very positive step in this direction. I trust you have found it to be beneficial also.